Lecture Notes in Mathematics

Edited by A. Dold and B. Eckmann

650

C*-Algebras and Applications to Physics

Proceedings, Second Japan-USA Seminar,
Los Angeles, April 18–22, 1977

Edited by H. Araki and R. V. Kadison

Springer-Verlag
Berlin Heidelberg New York 1978

Editors
Huzihiro Araki
Research Institute for
Mathematical Sciences
Kyoto University
Kyoto 606, Japan

Richard V. Kadison
Department of Mathematics E1
University of Pennsylvania
Philadelphia, PA 19104/USA

AMS Subject Classifications (1970): 81A15, 81A17, 82A15, 46K10, 46L05, 46L10

ISBN 3-540-08762-1 Springer-Verlag Berlin Heidelberg New York
ISBN 0-387-08762-1 Springer-Verlag New York Heidelberg Berlin

Printing and binding: Beltz Offsetdruck, Hemsbach/Bergstr.
2141/3140-543210

PREFACE

This volume contains the proceedings of the Second Japan-U.S. Seminar on C*-algebras and Applications to Physics. The seminar was sponsored jointly by the Japan Society for the Promotion of Science and the National Science Foundation (USA) who supplied travel and subsistence support for some of the participants. It was held at UCLA 18-22 April 1977.

The first five papers are extended versions of 90 minute talks presented during each of the five mornings of the seminar. They are expository accounts of broad and important areas of the subject. They appear in the order in which their authors spoke. The next group of shorter papers are by some of the speakers at the seminar. They appear in the order in which their authors spoke. The last three papers (and that of O. Takenouchi) are contributed by seminar participants. For the most part, these shorter papers describe detailed recent research. As such they offer a good opportunity to observe the methods and ideas of current interest in the subject.

A number of notes (of talks as well as other contributions) circulated at the seminar do not appear in this volume. They are primarily summaries of work appearing, or to appear, elsewhere. We list the authors and titles of these notes for the information this provides about the seminar.

Brown, L., Some Techniques in the Theory of C*-algebras

Bratteli, O., When is a C*-crossed Product Simple?

Bunce, J., A C*-algebraic Approach to Subnormal Operators

Feldman, J., Hahn, P. and Moore, C., Von Neumann Algebras and the Structure of Continuous Group Actions

Herman, R., Perturbations of Flows

Katayama, Y., Non-existence of a Normal Conditional Expection in a Continuous Crossed Product

Kishimoto, A., Equilibrium States of a Semi-quantum Lattice System

Nakagami, Y. and Sutherland, C., Takesaki's Duality for Regular Extensions of Von Neumann Algebras

Okayasu, T., Spectral Measures for *-automorphisms

Pedersen, G.K., An ABC on Spectral Theory for Groups of Automorphisms of Operator Algebras

Rieffel, M., Morita Equivalence for C*-algebras

__________, How Real Subspaces of Complex Hilbert Spaces are Related to Commutative Questions for Operator Algebras

Saito, K-S, On Non-commutative Hardy Spaces Associated with Flows on Finite Von Neumann Algebras

Takai, H., On the Invariant $\Gamma(\alpha)$ and C*-crossed Products

TABLE OF CONTENTS PAGE

ASPECTS OF NON-COMMUTATIVE ORDER

Edward G. Effros[1]

Notes for a lecture given at

The Second U. S. Japan Seminar on C*-algebras

and Applications to Physics

April 1977

1. Introduction

The simple notion that a C*-algebra is just "C(X) for a non-commutative X" has continued to prove quite fruitful. In recent years, we have witnessed the beginnings of both algebraic topology for "non-commutative spaces", specifically the theory of Ext, K_o, and K_1 for C*-algebras, and the non-commutative, or more precisely, matricial analogue of the theory of ordered vector spaces. The latter development has played a major role in the theory of operator algebras, and it is the subject that will be considered in these notes. Although, with the exception of §10, the material we consider has appeared elsewhere, we have given a fairly detailed account. We have chosen to do this because we feel that recent developments enable one to give a more coherent treatment than appears in the early papers. The results not attributed to others are largely due to work of Choi, Lance, and the author [12] - [17],[25].

The subject is somewhat technical. In order to simplify the exposition, we have relegated a number of definitions and simple results to an Appendix (§11). The reader will find that the latter can be largely ignored until § 7.

[1] Supported in part by NSF.

2. Scalar and matrix orderings

The underlying order and norm theoretic nature of the Banach space $C(X)$, the continuous real functions on a compact Hausdorff space X, has been understood for more than thirty-five years (see [1, p.78] and [33,§1] for a bibliography). In particular, given another space Y, Banach [6] and Stone [46] proved that if a unital linear map $C(X)\to C(Y)$ is either isometric or an order-isomorphism, then it is also an algebraic isomorphism. A decade later, Kadison discovered a remarkable non-commutative generalization of this result: if A and B are unital C^*-algebras, and A_h $\equiv Re(A)$ and B_h $\equiv Re(B)$ are the real subspaces of self-adjoint operators, then a unital isometric or order-isomorphic linear map $\varphi: A_h \to B_h$ must preserve the Jordan product

$$(a,b)\mapsto \tfrac{1}{2}(ab+ba).$$

The program suggested by Kadison's result was clear: one should attempt to study and ultimately to classify C^*-algebras A by examining the ordered Banach spaces A_h. This has proved to be a very difficult task. It was soon realized that the first step one must presumably take, the characterization of the spaces A_h, was equivalent to determining which compact convex sets can arise as the state spaces of unital C^*-algebras. For commutative C^*-algebras the answer is quite simple: the Choquet simplexes with closed extreme boundaries. In the general case, a complete answer to this question may finally be at hand, due to the work of Alfsen, Schultz, and Størmer [2]. In any event, Kadison's ordered Banach space approach, which I would like to call the scalar theory, has not yet fulfilled its initial promise.

In order to understand the obstacles to the scalar theory, and to anticipate the formulation of the matrix approach, let us consider a simple problem. Given a unital C^*-algebra A, the norm and order on A_h each determine the other. This is immediate from the fact that if $a \in A_h$,

$$(2.1)\qquad \|a\| \le 1 \iff -1 \le a \le 1.$$

On the other hand, the norm of a non-self-adjoint element in A would seem to have little relation to the ordering, which makes sense only on A_h. The way out of this difficulty is to note that the norm on A is in fact determined by the ordering on $M_2(A)$, the 2×2 matrices $[a_{ij}]$, $a_{ij} \in A$. This is a consequence of the fact that for <u>any</u> $a \in A$,

(2.2) $$\|a\| \leq 1 \iff \begin{bmatrix} 1 & a \\ a^* & 1 \end{bmatrix} \geq 0,$$

which is apparent from the formulas

$$\begin{bmatrix} 1 & a \\ a^* & 1 \end{bmatrix} = \begin{bmatrix} 1 & a \\ 0 & 0 \end{bmatrix}^* \begin{bmatrix} 1 & a \\ 0 & 0 \end{bmatrix} + \begin{bmatrix} 0 & 0 \\ 0 & 1-a^*a \end{bmatrix}$$

and

$$\begin{bmatrix} 0 & 0 \\ 0 & 1-a^*a \end{bmatrix} = \begin{bmatrix} 0 & a \\ 0 & -1 \end{bmatrix}^* \begin{bmatrix} 1 & a \\ a^* & 1 \end{bmatrix} \begin{bmatrix} 0 & a \\ 0 & -1 \end{bmatrix} .$$

It would thus seem advisable to regard the ordering on $M_2(A)$, and more generally on all of the matrix algebras $M_n(A)$, $n \geq 1$, as part of the natural "baggage" of a C*-algebra. The resulting category of "matrix ordered spaces" has proved to be of great value.

3. Completely positive maps

Following the historical development of the subject, we shall introduce the relevant morphisms, before carefully examining the objects to which they apply. In an attempt to clarify Nagy's theory of dilations, Stinespring [45] defined a linear map φ between two unital C*-algebras A and B to be completely positive if for all n, the corresponding maps

(3.1) $\qquad \varphi_n : M_n(A) \to M_n(B) : [a_{ij}] \mapsto [\varphi(a_{ij})]$

are positive. To see that this notion is more restrictive than positivity, we observe that although the transpose map on scalar 2×2 matrices M_2

$$\varphi : M_2 \to M_2 : \begin{bmatrix} \alpha & \beta \\ \gamma & \delta \end{bmatrix} \longmapsto \begin{bmatrix} \alpha & \gamma \\ \beta & \delta \end{bmatrix}$$

is positive (note that determinant and trace are preserved), it is not completely positive. To see this, note that if ε_{ij} are the matrix units in M_2,

$$\mathcal{E} = \begin{bmatrix} \varepsilon_{11} & \varepsilon_{12} \\ \varepsilon_{21} & \varepsilon_{22} \end{bmatrix} = \begin{bmatrix} \varepsilon_{11} & \varepsilon_{12} \\ 0 & 0 \end{bmatrix}^* \begin{bmatrix} \varepsilon_{11} & \varepsilon_{12} \\ 0 & 0 \end{bmatrix}$$

is positive (one has $M_2(M_2) \cong M_4$), whereas

$$\varphi_2(\mathcal{E}) = \begin{bmatrix} \varepsilon_{11} & \varepsilon_{21} \\ \varepsilon_{12} & \varepsilon_{22} \end{bmatrix} = \begin{bmatrix} 1 & 0 & 0 & 1 \\ 0 & 0 & 1 & 0 \\ 0 & 1 & 0 & 0 \\ 1 & 0 & 0 & 1 \end{bmatrix}$$

is not (examine the outlined 2×2 matrix.)

On the other hand, Stinespring proved that if A is a unital C*-algebra, states $\rho : A \to \mathbb{C}$ as well as unital *-homorphisms $\pi : A \to B$ are completely positive. More significantly, he proved that the classic GNS association of cyclic representations to states has a non-scalar generalization. Let us define an operator state of A on a Hilbert space K to be a completely positive unital map $\varphi : A \to \mathcal{B}(K)$. Stinespring proved that by using φ, one may construct a Hilbert space H, a *-representation $\pi : A \to \mathcal{B}(H)$, and a partial isometry $V : K \to H$ for which

(3.3) $\qquad \varphi(a) = V^* \pi(a) V$

and the projection $E = VV^*$ is cyclic in the sense that

$$[\varphi(A)EH]^- = H.$$

The most transparent argument for this result was given by Arveson [3], and closely resembles the GNS construction. The essential idea is to define a scalar product on $A \otimes H$ by

$$a \otimes x \cdot b \otimes y = (b^* a)x \cdot y,$$

and to let K be the associated Hilbert space. The representation π is induced by the following action of A on $A \otimes H$:

$$a(b \otimes x) = ab \otimes x.$$

The distinction between positivity and complete positivity is a non-commutative phenomenon: If A and B are unital C^*-algebras either one of which is commutative, then any unital positive map $A \to B$ is already completely positive [45],[3].

It is immediate from (3.3) that completely positive maps $\varphi: A \to B$ satisfy the Cauchy-Schwarz inequality

$$\varphi(a)^* \varphi(a) \leq \varphi(a^* a)$$

(a more difficult inequality was proved earlier by Kadison for positive maps [35].) What is more interesting is Choi's result [9] that if a is an element of A with

$$\varphi(a)^* \varphi(a) = \varphi(a^* a),$$

Then a is in the <u>left multiplicative domain</u> of φ, i.e.,

$$\varphi(b)\varphi(a) = \varphi(ba)$$

for all $b \in B$. From this, Choi proved that if $\varphi: A \to B$ is a unital matricial order isomorphism, then it is a *-isomorphism.[1] This analogue of Kadison's Theorem provides additional evidence that matricial rather than scalar theories might be more appropriate for studying non-commutative systems.

It was Arveson [3] who made what is the most fundamental observation concerning completely positive maps. In an attempt to formulate a theory of non-commutative function algebras, he found it essential to use operator states. He was then confronted with the functional analyst's imperative to extend. The Hahn-Banach Theorem and its order-theoretic version due to Krein were not applicable, hence Arveson was led to prove an analogue of these extension theorems for operator states. Before explaining this result and its relation to its antecedents, it is convenient to introduce the appropriate objects.

[1] It is my understanding that this was also known by E. Størmer.

4. The objects

We shall be comparing three categories in this paper:

$\mathcal{N}$: real normed vector spaces, contractions

$\mathcal{F}$: Kadison's function systems [33], positive unital maps

$\mathcal{O}$: operator systmes, completely positive unital maps

Rather than give the abstract definitions for the objects in $\mathcal{F}$ and $\mathcal{O}$, (see [33,§2],[12,§4]) we prefer to explain how they arise concretely.

To begin with, any real normed vector space V may be realized as a subspace of $C(X)$ with X a compact Hausdorff space. One way to do this is to let X be the morphisms $V \to \mathbb{R}$, i.e., the closed unit ball in V^* with the weak* topology, and to use the canonical injection $V \hookrightarrow C(X)$. Thus we regard subspaces of $C(X)$ as the concrete normed vector spaces.

A concrete function system is a subspace V of $C(X)$ that contains the constant function 1. The latter condition implies that the relative ordering on V is non-trivial, since if $v \in V$,

$$v = \|v\|1 - (\|v\|1 - v)$$

implies that $V = V^+ - V^+$. We define a function system to be any unital ordered vector space which is unitally order isomorphic to a concrete function system. It should be noted that (2.1) holds in a concrete function system, hence the norms obtained by the various realizations of a function system must coincide. The morphisms $\varphi : V \to \mathbb{R}$ are called states, and they comprise a weak* compact convex set $S \subseteq V^*$. We have the natural realization $V \hookrightarrow C(S)$.

A concrete operator system is a self-adjoint subspace V of $\mathcal{B}(H)$,

H a complex Hilbert space, containing the identiy operator 1. V is (non-trivially) matrix ordered, i.e., one has orderings on each of the spaces $M_n(V)$, because we have the inclusion

$$M_n(V) \subseteq M_n(\mathcal{B}(H)) \cong \mathcal{B}(H \oplus \ldots \oplus H).$$

An <u>operator system</u> is any unital matrix ordered space (see §11) which is unitally matrix order isomorphic to a concrete operator system. The norm on such a space is uniquely determined by (2.2). Generalizing our earlier definition, we define an <u>operator state</u> on V to be a morphism $\varphi : V \to \mathcal{B}(K)$ for some Hilbert space K. It is a simple exercise to prove that if V is an operator system, then the real subspace V_h of self-adjoint elements is a function system.

We may now formulate the Hahn-Banach Theorem for $\mathcal{h}$, the Krein variant for $\hat{\mathcal{f}}$, and the Arveson-Hahn-Banach Theorem for Θ/ simultaneously. Given $V \subseteq W$ in $\mathcal{h}$ (resp., $\hat{\mathcal{f}}$ or Θ - for these we always assume inclusions are unital), any contraction in V^* (resp., state or operator state on V) has an extension to such a map on W.

For the case that V is an operator system and K is finite dimensional, i.e., we have a unital completely positive map $\varphi : V \to M_n$, where n is the dimension of K, Arveson's Theorem can be reduced to the Krein Theorem by means of the following result (this is a special case of Lemma 11.1):

<u>Lemma 3.1</u>: Suppose that V is an operator system: The bounded completely positive maps $\varphi : V \to M_n$ can be identified with the positive functionals in $M_n(V)^*$. Given $p = [p_{ij}] \in M_n(V)^*$, the corresponding completely positive map is given by $a \mapsto [p_{ij}(a)]$.

Thus to extend φ, all we need to do is to extend the corresponding functional $p \in M_n(V)^{*+}$ to a functional $q \in M_n(W)^{*+}$. Restricting p to the function system $M_n(V)_h$, the Krein Theorem provides the self-adjoint restriction of q (a form of this argument was shown to us some years ago by G. Elliott - it also occurs in [47]).

Rather than prove Arveson's Theorem for infinite dimensional operator states, we turn to a more general problem.

5. Injectives and amenable groups

In any of the categories $\mathcal{N}$, $\mathcal{F}$, $\mathcal{O}$ we may define an object R to be <u>injective</u> if given $V \subseteq W$, any morphism $\varphi: V \to R$ may be extended to a morphism $\psi: W \to R$. Thus the Hahn-Banach Theorem and the Krein and Arveson variants simply state that in the appropriate categories, $\mathbb{R}$ and $\mathcal{B}(H)$ are injective. The characterization of the injectives in $\mathcal{N}$ by Nachbin, Goodner, and Kelley ([42], [28], [36] - see [5] for an elegant presentation) represented the first basic classification result in the theory of Banach spaces. The theory for $\mathcal{F}$ closely parallels that for $\mathcal{N}$. On the other hand, the study of injectives in $\mathcal{O}$, culminating in the determination of the injective von Neumann algebras by Connes [19], must be regarded as the greatest triumph in the structure theory of von Neumann algebras since the work of Murray and von Neumann.

The theory of injectives for $\mathcal{N}$ (resp., $\mathcal{F}$) is elegant and simple: a normed vector space V is injective if and only if it is isometric (resp., unitally order isomorphic) to $C(X)$, where X is a Stonean space. If V is a dual Banach space in either $\mathcal{N}$ or $\mathcal{F}$, we may replace $C(X)$ by $L^\infty(X, \mathcal{S}, \mu)$ where $(X, \mathcal{S}, \mu)$ is a measure space. An immediate, but surprisingly important technical consequence of these results is that since $C(X)^{**} \cong C(Y)$ for some Stonean space Y, if V is injective, the same is true for V^{**}.

Turning to $\mathcal{O}$, once again one has that the injectives (resp., dual injectives) must be isomorphic to C*-algebras (resp, von Neumann algebras) [12]. The first clue that a much more profound outlook is required in the theory for $\mathcal{O}$ is Hakeda and Tomiyama's observation [31] that the converse is false: there exist non-injective von Neumann algebras. Their counter-examples were motivated by an argument of J. Schwartz [44], who showed that there exist von Neumann algebras which do not have the closely related "Property P". Quickly summarizing the argument, let G be a countable discrete group and let $R(G)$ be the von Neumann algebra on $H = l^2(G)$ determined by left translation

$$V(s): H \to H : f \mapsto sf, \qquad f \in l^2(G),$$

where for any function f on G we let

$$sf(t) = f(s^{-1}t).$$

For each $s \in G$, we let δ_s be the corresponding element of H, i.e.,

$$\delta_s(t) = \begin{cases} 0 & t \neq s \\ 1 & t = s \end{cases},$$

and we define a trace on $R(G)$ by

$$\tau(a) = a\delta_e \cdot \delta_e ,$$

where e is the identity element of G. Letting

$$M: l^\infty(G) \to \mathcal{B}(H)$$

be the multiplication representation, we have that

$$V(s)M(h)V(s^{-1}) = M(sh), \qquad h \in l^\infty(G).$$

Assuming that $R(G)$ it is injective, it follows that the identity map $R(G) \to R(G)$ has a completely positive extension $P: \mathcal{B}(H) \to R(G)$. Since $R(G)$ lies in the left and right multiplicative domains of P (see §3), we conclude that

$$\mu : l^\infty(G) \to \mathbb{C} : h \mapsto \tau(P(h))$$

satisfies

$$\mu(sh) = \tau(V(s)^* P(M(h))V(s)) = \mu(h),$$

i.e., μ is an invariant mean on G. Thus we have associated a non-injective von Neumann algebra $R(G)$ to any non-amenable group G such as $G = \mathbb{F}_2$, the free group on two generators.

Roughly speaking, a group is amenable if and only if it has subsets which suffer only a small relative movement under a given compact set of translations. For discrete groups, the precise condition is due to Følner (see [29, p.64]): given $K \subseteq G$ finite, and $\varepsilon > 0$, there exists a set $\emptyset \neq U \neq G$ such that

$$\frac{|(sU) \Delta U|}{|U|} < \varepsilon, \qquad s \in K,$$

where $|\ |$ indicates cardinality, and Δ symmetric difference. Discrete groups determine finite von Neumann algebras, and the first step in Connes' theory of injectives is to prove that the finite factors are characterized by an analogue of the Følner Condition (Wasserman has recently shown that essentially the same arguments apply to global finite algebras [50]). The latter states that given a finite set K in a finite factor R, and an $\varepsilon > 0$, there exists a finite dimensional projection $e \in \mathcal{B}(H)$ such that for all $a \in K$,

$$\|[a,e]\|_{HS} \leq \varepsilon \|e\|_{HS}$$

and

$$|\tau(a_j) - \langle a_j e, e\rangle_{HS} / \langle e,e\rangle_{HS}| < \varepsilon$$

where $\langle\ ,\ \rangle_{HS}$ and $\|\ \|_{HS}$ are the Hilbert-Schmidt scalar product and norm, respectively. By using a remarkable sequence of steps (we have left out many of them), Connes proved that if R is a finite factor on a separable Hilbert space,

R injective $\Rightarrow$ R satisfies the analogue of the Følner condition
$\Rightarrow$ the identity map $R \to R$ is semidiscrete
$\Leftrightarrow$ R is semidiscrete
$\Rightarrow$ R is the weak closure of an ascending sequence of matrix algebras (i.e., R is hyperfinite)
$\Rightarrow$ R is injective.

He then went on to prove that these are all equivalent for arbitrary factors on a separable Hilbert space by using the decomposition theory that he and Takesaki had earlier developed. The notion of semidiscreteness is directly concerned with the matrix ordering of a von Neumann algebra, and will be considered in §8.

The first implication is proved by using a non-commutative form of Day's "convergence to invariance" argument [29], including a non-commutative L^p inequality due to Powers and Størmer. The second implication is straightforward and depends upon the properties of the finite-dimensional projections of the Følner condition. Connes uses both parts of the equivalence, which is proved in §8. The following implication is by far the most remarkable, using the automorphism groups and the asymptotic properties of R. The final implication is easy. The equivalence of these results for an arbitrary von Neumann algebra on a separable Hilbert space is undoubtedly valid, but we have not seen an explicit proof of hyperfiniteness in this context. All of the other conditions are equivalent for von Neumann algebras on arbitrary Hilbert spaces [50].

The great importance of Connes' Theorem is that it is much easier to prove injectivity than hyperfiniteness. Thus with his result we now have an affirmative solution to the Kadison-Singer conjecture: if a locally compact second countable group G is either solvable or connected, then all of its factor representations generate hyperfinite factors.

One more remark is relevant to this section. Since von Neumann algebras need not be injective, we cannot conclude that R injective implies the same is true for R^{**}. Indeed, Wasserman [49] has proved that $\mathcal{B}(H)^{**}$ is not injective. We shall return to this fact in §10.

6. C*-algebras with injective representations

Von Neumann algebras generally arise from representations of C*-algebras, hence it is natural to ask when a given C*-algebra can generate only von Neumann algebras of a specified kind. For example, since a connected Lie group which is semisimple or nilpotent can have only type I representations (see [32],[39]), the same is true for the corresponding full group C*-algebras. C*-algebras with this property are said to be of type I, and owing to the results of Dixmier [20], Glimm [27], and Sakai [43], they have been characterized in a number of ways. It follows from the previous section that the full group C*-algebra of any connected Lie group has only injective representations, hence we are confronted with the problem of determining which C*-algebras have only injective representations.

Since we are concerned with all of the representations of a given C*-algebra A, we are led to investigate the universal representation of A, or equivalently, the enveloping von Neumann algebra A^{**} (see [21,§XII]). It is not difficult to show that the algebras A that we seek to characterize are just those for which A^{**} is injective.

Turning to the categories $\mathfrak{L}$ and $\mathfrak{F}$ for guidance, we find that there is already extant a number of results about the objects V with V^{**} injective. In $\mathfrak{L}$ these are just the "Lindenstrauss spaces" (see, e.g., [38]), whereas in $\mathfrak{F}$ they are the "simplex spaces" (see [23]). Restricting our attention to $\mathfrak{F}$ we have (see [40])

<u>Theorem 6.1</u>: If V is a function system, the following are equivalent:

(1) V^{**} is injective (i.e., V is a simplex space).

(2) V is nuclear, i.e., the $\mathfrak{F}$ minimal and maximal tensor products of V with any function system must coincide.

(3) (V separable) V is an inductive limit of finite dimensional injective function spaces (these must be the l^{∞} spaces of finite point sets).

The third condition is, of course, the most interesting one, and may be regarded as a major step in the classification of simplex spaces. Choi [11] has recently shown that the strict analogue of (3) is false even for type I C*-algebras. On the other hand it would seem likely that some structural analogue of Glimm's "type I $\Leftrightarrow$ GCR" exists for the C*-algebras satisfying (1). We feel that the formulation and proof of such a result is the central problem today in the algebraic theory of C*-algebras.

The equivalence (1)$\Leftrightarrow$(2) is valid for C*-algebras. Before introducing the appropriate notion of nuclearity, we must consider the general theory of tensor products of matrix ordered spaces. The latter has proved to be a very powerful tool.

7. Tensor products and their relation to complete positivity

The various tensor products of C*-algebras arise quite naturally in representation theory. For example, suppose that G_1 and G_2 are countable groups. Then for the full group C*-algebras, we have

$$C^*(G_1 \times G_2) \cong C^*(G_1) \otimes_{max} C^*(G_2),$$

whereas for the regular group C*-algebras,

$$C^*_{reg}(G_1 \times G_2) \cong C^*_{reg}(G_1) \otimes_{min} C^*_{reg}(G_2).$$

The study of C*-algebras is greatly complicated when the minimal and maximal tensor products do not coincide. This is due to the fact that given C*-algebras A_1, A_2, and B, an injection $A_1 \hookrightarrow A_2$ induces an injection

(7.1) $$A_1 \otimes_{min} B \hookrightarrow A_2 \otimes_{min} B,$$

but the corresponding statement for $\otimes_{max}$ need not be true. On the other hand, if J is an ideal in A_1 and $A_2 = A_1/J$, one has that

(7.2) $$\ker (A_1 \otimes_{max} B \twoheadrightarrow A_2 \otimes_{max} B) = \overline{J \otimes B} \quad (\cong J \otimes_{max} B),$$

but one can have that

$$\ker (A_1 \otimes_{min} B \twoheadrightarrow A_2 \otimes_{min} B) \supsetneqq \overline{J \otimes B} \quad (\cong J \otimes_{min} B)$$

(see [30],[49]). These difficulties simultaneously vanish if the tensor products coincide. Using this fact and the solution to the Tomiyama problem (see §8), Blackadoor [7] proved that if B is nuclear, then for the primitive ideal spaces,

$$\text{pr } A \otimes_{min} B \cong \text{pr } A \times \text{pr } B.$$

The philosophy was advanced in [22 ; 24; 60] that the primitive ideal space is more appropriate for harmonic analysis on non-type I groups than Mackey's dual. Since this has since proved to be the case (see [41]), hence relations such as (7.3) are of great interest.

There is another more abstract reason for studying tensor products. The suffix "al" is a reminder of the fact that functional analysis was to begin with concerned with scalar maps. Tensor products have been useful since with them one can often replace vector-valued maps by scalar ones (this is really what underlies the proof of the finite Arveson-Hahn-Banach Theorem in §3). The algebraic reason for this is the isomorphism

$$(7.4) \qquad \mathrm{Lin}\,(V,W^d) \cong (V\otimes W)^d$$

in the category of vector spaces and linear maps. (7.4) is determined by the inverse maps $\varphi \mapsto f_\varphi$ and $f \mapsto \varphi_f$, where

$$(7.5) \qquad \begin{aligned} f_\varphi(v\otimes w) &= \varphi(v)(w) \\ \varphi_f(v)(w) &= f(v\otimes w) \end{aligned}$$

The basic observation made by Lance in [37] is that if A and B are C*-algebras, and we let $\mathcal{B}(A,B^*)$ have the ordering defined by the cone of completely positive maps, then we have an order isomorphism

$$(7.6) \qquad \mathcal{B}(A,B^*) \cong (A\otimes_{\min} B)^*.$$

This was the first indication that complete positivity and C*-algebraic tensor products were related, and it enabled Lance to make important progress in the study of nuclearity. Rather than prove Lance's result, we shall consider it in the more general context of matrix ordered spaces. The terminology and notation that we use are explained in §11.

Given dual matrix ordered spaces V, V^δ and W, W^δ, (7.4) restricts to the real linear isomorphism

(7.7) $\mathcal{B}(V,W^\delta)_h \cong (V\odot W)^\delta$.

We may order $\mathcal{B}(V,W^\delta)_h$ by the cone $\mathcal{B}(V,W^\delta)^+$ of completely positive maps. We are then faced with the question: For what ordering on $V\odot W$ is it the case that (7.7) is an order isomorphism? Given $f\in(V\odot W)^\delta$ we have that $\varphi_f:V\to W$ is completely positive if and only if for each n and $v\in M_n(V)^+$, $(\varphi_f)_n(v)\in M_n(W)^+$, i.e., given $v\in M_n(V)^+$ and $w\in M_n(W)^+$,

$$\begin{aligned} 0\leq(\varphi_f)_n(v)(w) &= [\varphi_f(v_{ij})][w_{ij}] \\ &= \sum\varphi_f(v_{ij})(w_{ij}) \\ &= \sum f(v_{ij}\otimes w_{ij}). \end{aligned}$$

Introducing the notation $v\times w=\sum v_{ij}\otimes w_{ij}$, we conclude that we should have that $f\geq 0$ if and only if $f(v\times w)\geq 0$ for all such v, w. Due to MIII in §11, the set

$$(V\odot W)^+=\{v\times w:\ v\in M_n(V)^+,\ w\in M_n(W)^+\}$$

is convex (see [12, p. 180]). It is evident that the appropriate cone in $V\odot W$ is

$$(V\odot_M W)^+ = \mathrm{Cl}\,(V\odot W)^+,$$

where we are using the duality between $V\odot W$ and $(V\odot W)^\delta$. We define the maximal tensor product (relative to the given dualities) $V\odot_M W$ to be $V\odot W$ with this ordering (this conflicts with [16,§4.4] unless V or W is finite-dimensional. The latter was the case throughout that paper).

We may instead start off with the cone $(V^\delta\odot W^\delta)^+$ inside $(V\odot W)^\delta$. Letting $(V\odot_m W)^+$ be the dual cone in $V\odot W$, we define the minimal tensor product $V\odot_m W$ to be $V\odot W$ with this cone. We will have that $V\odot_m W=V\odot_M W$ if and only if their dual cones in $(V\odot W)^\delta$ coincide, i.e.,

(7.8) $(V\odot_M W)^{\delta+} = \mathrm{Cl}(V^\delta\odot W^\delta)^+$.

Returning to (7.7), we must show that each completely positive map $\varphi: V \to W$ is defined by an element of the right side of (7.8). We must therefore ask the general question, *which maps $\varphi: V \to W^{\delta}$ are of the form φ_f, where $f \in Cl(V^{\delta} \odot W^{\delta})^{+}$*?

First let us suppose that $f \in V^{\delta} \odot W^{\delta}$ is of the form $f = g \times h$, $g \in M_n(V^{*})$ $h \in M_n(W^{\delta})$. Given $v \in V$ and $w \in W$ we have (see §11)

$$\begin{aligned} \varphi_f(v)\cdot w &= f\cdot(v \otimes w) \\ &= (\sum g_{ij} \otimes h_{ij})\cdot(v \otimes w) \\ &= (\sum g_{ij}(v)\, h_{ij})\cdot w \\ &= \Theta(h)\,[g_{ij}(v)]\cdot w \quad , \\ &= \Theta(h)\, \Lambda(g)(v)\cdot w \end{aligned}$$

i.e., $\varphi_f = \Theta(h)\Lambda(g)$. From Lemma 11.1, we have that $g \geq 0$ and $h \geq 0$ if and only if $\Lambda(g)$ and $\Theta(h)$ are completely positive. We conclude that $f \in (V^{\delta} \odot W^{\delta})^{+}$ if and only if for some n one has a commutative diagram

(7.9)
$$\begin{array}{ccc} & M_n & \\ {\scriptstyle \sigma} \nearrow & & \searrow {\scriptstyle \tau} \\ V & \xrightarrow[\varphi]{} & W \end{array}$$

where $\varphi = \varphi_f$, and σ, τ are completely positive. We say that such a map is *exactly nuclear*. If $\varphi = \varphi_f$ is exactly nuclear, it is of course completely positive, i.e., $f \in (V \odot W)^{\delta +}$, and we conclude

$$(V^{\delta} \odot W^{\delta})^{+} \subseteq (V \odot W)^{\delta +}.$$

Finally letting $\mathcal{B}(V, W^{\delta})$ have the point-weak topology, (7.7) is a homeomorphism and we have that $f \in Cl(V^{\delta} \odot W^{\delta})^{+}$ if and only if φ is a limit of exactly nuclear maps. A convenient way of stating this is to say that the diagrams (7.9) (with n arbitrary) *approximately commute*. If this is the case, we say that φ is *duality nuclear*. We conclude (see [37] for an earlier form of this result:

Theorem 7.1: Given dual matrix ordered spaces V, $V^{\$}$ and W, $W^{\$}$, we have that

$$V \odot_m W = V \odot_M W$$

if and only if every completely positive map $\varphi : V \to W^{\$}$ is duality nuclear.

In the context of C*-algebras and von Neumann algebras, this leads us to the notions of nuclearity and semidiscreteness, respectively.

8. Nuclearity and semidiscreteness

We may regard a C*-algebra A and its Banach dual A^* as dual matrix ordered spaces. Given two such algebras A and B, we then have that

$$A \odot_M B = (A \otimes_{max} B) \cap (A \otimes B)_h$$
$$A \odot_m B = (A \otimes_{min} B) \cap (A \otimes B)_h$$

where $A \otimes_{max} B$ and $A \otimes_{min} B$ are the maximal and minimal C*-algebraic tensor products. The latter are defined by completing the *-algebra $A \otimes B$ with respect to the norms

$$(8.1) \quad \begin{aligned} \|u\|_{max} &= \sup\{p(u^*u)^{\frac{1}{2}} : p \in (A \odot_M B)^{\delta +}, \ \|p\|_0 \leq 1\} \\ \|u\|_{min} &= \sup\{p(u^*u)^{\frac{1}{2}} : p \in (A^\delta \odot B^\delta)^{+}, \ \|p\|_0 \leq 1\} \end{aligned}$$

where $\|p\|_0$ is the usual bilinear norm

$$\|p\|_0 = \sup\{p(a \otimes b) : \|a\|, \|b\| \leq 1\}$$

(if A and B are unital, one may replace $\|p\|_0 \leq 1$ by $p(1 \otimes 1) = 1$). Given faithful representations $A \hookrightarrow \mathcal{B}(H)$, $B \hookrightarrow \mathcal{B}(K)$, $A \odot_{min} B$ may also be realized as the completion of $A \odot B$ in $\mathcal{B}(H \bar{\otimes} K)$.

We define a C*-algebra A to be nuclear if for all C*-algebras B, $A \otimes_{min} B = A \otimes_{max} B$, or equivalently, $A \odot_m B = A \odot_M B$. We say that a completely positive map $\varphi : A \to B$ is nuclear if the diagrams

$$\begin{array}{ccc} & M_n & \\ {}^{\sigma}\nearrow & & \searrow{}^{\tau} \\ A & \xrightarrow[\varphi]{} & B \end{array} \qquad \text{(n arbitrary)}$$

with σ, τ completely positive contractions, approximately commute in the point-norm topology. The terminology is explained by

Theorem 8.1 [13]: A C*-algebra A is nuclear if and only if $id : A \to A$ is nuclear.

This result provides the key step in proving the Completely Positive Lifting Theorem for separable nuclear C*-algebras [16] (this is still the case for the recent proofs [52], [53]). The latter is used for showing that Ext of a separable nuclear C*-algebra is a group, and is known to be false for separable non-nuclear C*-algebras [17] (in fact, Ext need not be a group [51]).

It is instructive to sketch the argument for Theorem 8.1. One first shows that it suffices to assume that A is unital. Given any von Neumann algebra R, and using the dualities A, A^* and R, R^*, we have that $A \otimes_m R = A \otimes_M R$. This implies the same is true if we use the R, R_* duality (the cone on the left is unaffected, that on the right lies between the previous two cones). From Theorem 7.1, completely positive maps $A \to R_*$ must be duality nuclear. The usual convexity argument enables one to replace point-weak with point-norm convergence. If $\varphi \neq 0$ is such a map, $p = \varphi(1)$ is a non-zero positive normal linear functional on R, and thus defines a normal representation π_p of R. We have that $\varphi(A) \subseteq [p]$, where $[p]$ denotes the subspace of R_* generated by the cone $\{q:\ 0 \leq q \leq \alpha p, \alpha > 0\}$. On the other hand, we have a complete order isomorphism $[p] \cong \pi_p(R)'$ (this is just the usual correspondence that shows, for example, the relation between pure states and irreducible representations). In fact by using Sakai's Radon Nykodym Theorem, one finds that any completely positive map $\varphi: A \to \pi_p(R)'$ is suitably nuclear. By juggling the von Neumann algebras R, we can arrange to have that the same is true for the inclusion $A \hookrightarrow A^{**}$. More precisely, we have that the diagrams

$$(8.2) \qquad \begin{array}{ccc} & M_n & \\ \sigma \nearrow & & \searrow \tau \\ A & \hookrightarrow & A^{**} \end{array}$$

approximately commute in the point σ-weak topology. But we can use the order isomorphism

$$\mathcal{B}(M_n, A^{**}) \cong M_n(A^{**}) = M_n(A)^{**}$$

(Lemma 11.1), together with the fact that $M_n(A)^+$ is weakly dense in $M_n(A)^{**+}$, to approximate $\tau: M_n \to A^{**}$ by maps $\tau': M_n \to A$. Convexity again enables one to use the point-norm topology.

If we regard a von Neumann algebra R and its predual R_* as dual matrix ordered spaces, we obtain possibly distinct tensor products. In fact, using this duality for von Neumann algebras R and S, we have

$$R \odot_M S = (R \otimes_{bin} S) \cap (R \otimes S)_h$$

$$R \odot_m S = (R \otimes_{min} S) \cap (R \otimes S)_h$$

where $R \otimes_{bin} S$ is the binormal tensor product obtained by completing the $*$-algebra $R \otimes S$ with respect to the norm

$$\|u\|_{bin} = \sup \{ p(u^*u)^{\frac{1}{2}} : p \in (R \odot_M S)^{\delta +}, \ \|p\|_0 \leq 1 \} \quad (R, R_* \text{ duality})$$

and $R \otimes_{min} S$ is just the minimal tensor product defined above. It should be stressed that we are dealing with C*-algebraic completions of $R \otimes S$.

A von Neumann algebra R is said to be semidiscrete [25] provided for all von Neumann algebras S,

$$R \otimes_{bin} S = R \otimes_{min} S.$$

In order to understand the importance of this concept, we recall that if $R \subseteq \mathcal{B}(H)$ is a factor, the map

(8.3) $\quad R \otimes R' \to RR' : r \otimes r' \mapsto rr'$

is an algebraic $*$-isomorphism. We may regard $R \otimes R'$ as a $*$-algebra of operators on $H \otimes_2 H$. (8.3) is almost never σ-weakly continuous, and thus does not extend to the σ-weak closure $R \bar{\otimes} R'$ (i.e., the von Neumann algebra tensor product). On the other hand it is norm continuous with respect to $\| \ \|_{bin}$, and thus extends to $R \otimes_{bin} S$. It follows that if R is semidiscrete, then (8.3) is norm continuous relative to the norm on $\mathcal{B}(H \otimes_2 H)$, since the latter determines the minimal norm Since R is a factor, (8.3) is in fact isometric. This property for semidiscrete factors plays an important role in Connes' theory.

Given von Neumann algebras R and S, we say that a map $\varphi : R \to S$ is <u>semidiscrete</u> provided the diagrams

$$\begin{array}{ccc} & M_n & \\ \sigma \nearrow & & \searrow \tau \\ R & \xrightarrow[\varphi]{} & S \end{array}$$

with σ, τ completely positive σ-weakly continuous contractions, approximately commute in the point σ-weak topology. An argument similar to that sketched for Theorem 8.1 gives

<u>Theorem 8.2</u> [25],[14]: A von Neumann algebra R is semidiscrete if and only if $id: R \to R$ is semidiscrete.

The definitive result for semidiscrete von Neumann algebras is:

<u>Theorem 8.3</u>: Suppose that R is a von Neumann algebra. Then the following are equivalent:

(1) R is semidiscrete

(2) the map $R \otimes R' \to RR'$ is norm-decreasing ($R \otimes R' \subseteq \mathcal{B}(H \otimes_{\wedge} H)$)

(3) R is injective.

The proof that (1)$\Leftrightarrow$(2)$\Rightarrow$(3) may be found in [25], and we shall sketch a simple argument for (1)$\Rightarrow$(3) in the next section. The implication (3)$\Rightarrow$(1) was discussed in §5.

As indicated in §6, the notions of nuclearity and injectivity are closely related. In fact, we have

<u>Theorem 8.4</u>: If A is any C*-algebra, then the following are equivalent:

(1) A^{**} is semidiscrete

(2) A is nuclear

(3) A^{**} is injective

The implications (1) $\Rightarrow$ (2) $\Leftrightarrow$ (3) were proved in [25],[15]. The full equivalence is obtained by replacing (3) $\Rightarrow$ (2) by (3) $\Rightarrow$ (1) (Theorem 8.3). To see that (1) $\Rightarrow$ (2), note that given approximately commutative diagrams

$$\begin{array}{ccc} & M_n & \\ \nearrow & & \searrow \\ A^{**} & \xrightarrow[id]{} & A^{**}, \end{array}$$

restriction to A gives the diagrams (8.2), and as before, this implies that A is nuclear. An argument for (2) $\Rightarrow$ (3) may be found in §9. One might think that (2) $\Rightarrow$ (1) is immediate since all one need do is take second adjoints of the diagrams

$$\begin{array}{ccc} & M_n & \\ \nearrow & & \searrow \\ A & \xrightarrow[id]{} & A \end{array} .$$

This argument does not work since one cannot prove convergence on elements in $A^{**} \setminus A$. We regard Theorem 8.4 as one of the deepest results in the subject. Perhaps the most important consequence of (2) $\Leftrightarrow$ (3) is the solution of <u>Tomiyama's problem</u>: ideals, quotients, and extensions preserve nuclearity [14][15].[1]

"Hybrid" tensor products can be obtained between von Neumann algebras and C*-algebras by using the Banach duals and preduals, respectively. This idea essentially occurred in our proof of Theorem 8.1, and is further discussed in [25].

---

[1] I am indebted to A. Wulfsohn for pointing out to me that one half of Tomiyama's problem is actually quite simple, and was already known. To be specific, T. Huruya proved that A is nuclear if that is true for a closed two sided ideal J and the quotient A/J. The relevant reference (to which I have not had access) is [61]. . The converse implication remains very difficult.

9. A return to injectivity

Many characterizations exist for the injectives in $\mathcal{\lambda}$ and $\mathcal{F}$. Perhaps the most interesting of these are concerned with the internal structure of the spaces. Thus a real normed vector space is injective if and only if any collection of closed balls that intersect two at a time has a simultaneous intersection (see [5]). On the other hand a function system V is injective if and only if it is lattice ordered and conditionally order complete. More generally (recall that in $\mathcal{\lambda}$ and $\mathcal{F}$, V injective $\Rightarrow$ V^{**} injective) a space V in $\mathcal{\lambda}$ (resp., $\mathcal{F}$) is such that V^{**} is injective if and only if the intersection property holds for finite collection of balls (resp., V satisfies the Riesz decomposition property). The latter result can be proved by considering certain tensor products of V with canonical finite-dimensional spaces (see [40],[55],[56], and §10). Analogous techniques have led to interesting results for $\mathcal{O}$, which we shall briefly explain.

For each $n \geq 1$, we define N_{2n} to be the operator system

$$N_{2n} = \{\alpha \in M_{2n} : \sum_{i \leq n} \alpha_{ii} = \sum_{i > n} \alpha_{ii}\}$$

We then have (see [15],[12] for a slightly weaker result):

Theorem 9.1: If R is a von Neumann algebra, then the following are equivalent:

(1) R is injective

(2) any completely positive map $N_{2n} \to R$ has a completely positive extension to M_{2n}

(3) any completely positive map $N_{2n} \to R_*$ has a completely positive extension to M_{2n}

The proof of the equivalence (1)$\Leftrightarrow$(2) depends upon a careful analysis of what is required to extend a completely positive map from an operator system N into R to $N + \mathbb{C}a$, where a lies in a larger operator system. (2)$\Leftrightarrow$(3) is based upon the fact that R_* may be thought of as an assemblage of the von Neumann algebras $\pi_p(R)'$, $p \in R_*$ (see the discussion of Theorem 8.1) and that $\pi_p(R)'$ is injective if and only if that is the case for $\pi_p(R)$ ([57] - ultimately a consequence of the Tomita-Takesaki theory).

The significance of (2) and (3) is that they have important tensor product interpretations. Thus using the order isomorphism (7.5), (2) is equivalent to the assertion that the restriction map

$$(M_{2n} \odot_M R_*)^{\delta} \to (N_{2n} \odot_M R_*)^{\delta}$$

is order surjective, i.e., the positive cone on the left is mapped onto the positive cone on the right. Using some technical results from the theory of ordered vector spaces (certain conditions must be checked - see 18, Vol.II) we find that it is equivalent to assume that

$$(9.1) \qquad N_{2n} \odot_M R_* \hookrightarrow M_{2n} \odot_M R_*$$

is an order injection, i.e., if $u \in N_{2n} \odot R_*$ is positive as an element of $M_{2n} \odot_M R_*$, then it is already positive in $N_{2n} \odot_M R_*$.

For any finite dimensional matrix ordered space N one has the linear isomorphism

$$N \odot R_* \cong (N^d \odot R)^{\delta} \cong \mathcal{B}(R,N).$$

Since $(N \odot_m R_*)^+$ and $(N^d \odot R)^{\delta +}$ are both defined as dual to $(N^d \odot R)^+$ (if N had been infinite dimensional, these would be in different spaces), they coincide, i.e.,

$$(N \odot_m R_*)^+ \cong \text{all completely positive maps } R \to N.$$

On the other hand

$$(N \odot_M R_*)^+ \cong \text{all duality nuclear maps } R \to N.$$

Of course if $N \equiv M_{2n}$, any completely positive map $R \to N$ is automatically exactly nuclear, i.e., we have from Lemma 11.1

$$M_{2n} \odot_M R_* \cong \mathcal{B}(R,\, M_{2n}) \cong M_{2n}(R)_*.$$

Reinterpreting the order injectivity of (9.1), we first note that the maps $R \to N_{2n}$ which are completely positive as maps $R \to M_{2n}$, are simply the completely positive maps $R \to N_{2n}$. Thus (2) is equivalent to stating that any completely positive map $R \to N_{2n}$ is duality nuclear. Applying a similar argument to (3), we have

Theorem 9.2 [12]: If R is a von Neumann algebra, then the following are equivalent:

(1) R is injective

(2) any completely positive map $R \to N_{2n}$ is duality nuclear

(2') $N_{2n} \odot_m R_* = N_{2n} \odot_M R_*$ for all n

(3) any completely positive map $N^d_{2n} \to R$ is duality nuclear

(3') $N_{2n} \odot_m R = N_{2n} \odot_M R$.

From either (2) or (3) it is immediate that

$$R \text{ semidiscrete} \Rightarrow R \text{ injective}$$

because using (3), for example, we need only consider the diagrams

$$N^d_{2n} \to R \begin{array}{c} \nearrow M_k \searrow \\ \xrightarrow[\;id\;]{\qquad\qquad} \end{array} R.$$

An analogous theory holds in part for C*-algebras A. Using the duality A, A^*, we have

Theorem 9.3 [12]: If A is a unital C*-algebra, then the following are equivalent:

(a) A^{**} is injective

(b) any completely positive map $N_{2n}^{d} \to A$ is duality nuclear

(b') $N_{2n} \odot_m A = N_{2n} \odot_M A$ for all n.

From (b) it is immediate that

$$A \text{ nuclear} \Rightarrow A^{**} \text{ injective}$$

since we have the diagrams

$$N_{2n}^{d} \to A \begin{array}{c} \nearrow M_k \searrow \\ \xrightarrow[\text{id}]{\qquad\qquad} \end{array} A \;.$$

It came as quite a surprise that the analogues of (2) and (2') in Theorem 9.2 are false for C^*-algebras. Thus $\mathcal{B}(H)$ (regarded as a C^*-algebra) satisfies that condition [12, Cor. 6.3] but as we have previously remarked, Wasserman has shown $\mathcal{B}(H)^{**}$ is not injective.

10. A conjecture about some canonical bad apples

From Theorems 8.4 and 9.3 we have

$$A \text{ is nuclear} \Leftrightarrow A^{**} \text{ is injective}$$
$$\Leftrightarrow A \otimes_m N_{2n} = A \otimes_M N_{2n} \qquad \text{(all n).}$$

This is the analogue of a beautiful result of Namioka and Phelps [40] in the category $\mathfrak{F}$. Let $\mathbb{R}^{2n}$ have the ordering

$$(\mathbb{R}^{2n})^+ = \{\alpha : \alpha_i \geq 0, \quad 1 \leq i \leq 2n\}$$

and let

$$F_{2n} = \{\alpha \in \mathbb{R}^{2n} : \sum_{i \leq n} \alpha_i = \sum_{i > n} \alpha_i\}$$

have the relative ordering. As ordered vector spaces, we may identify $\mathbb{R}^{2n}$ with $C(X)$, where X consists of 2n points. If $n = 2$, F_4 is order isomorphic to the function system of affine functions on a square. Using suitable function system tensor products, we have

Theorem 10.1: If V is a function system, then

$$V \text{ is nuclear} \Leftrightarrow V^{**} \text{ is injective}$$
$$\Leftrightarrow V \otimes_m F_4 = V \otimes_M F_4$$

Regarding F_4 as a function system, it is of course not nuçlear since its state space, the square, is not a simplex. Another way to see this is to note that it is not lattice ordered (for finite dimensional spaces this is equivalent to the Riesz decomposition property), since cross-sections of F_4^+ are squares rather than simplexes.(see [18, Vol. II]).

The matrix ordered spaces N_{2n}^d would thus seem to be matricial analogues of certain non-simplicial polyhedra. Some information is known about them. They are completely order isomorphic to operator systems [12,§4]. On the other hand, N_2^d cannot be realized as an operator system on a finite dimensional Hilbert space [12,§7.2], and in particular, cannot be matricially order isomorphic to N_2. This is in sharp contrast to F_4^d, which is order isomorphic to F_4, and thus can be realized as a function on a four point set. More surprisingly, we have

Theorem 10.2: There exists an n_o such that $N_{2n_o}^d$ cannot be matricially order isomorphic to an operator system inside a nuclear C*-algebra.

Proof: Since $\mathcal{B}(H)^{**}$ is not injective, we have from Theorem 9.3 (b) that there exists an n and a completely positive map $\varphi : N_{2n_o}^d \to \mathcal{B}(H)$ which is not nuclear in the point-norm sense. If one had $N_{2n_o}^d \subseteq A$, with A nuclear, we could use injectivity of $\mathcal{B}(H)$ to extend φ to a completely positive map $\psi : A \to \mathcal{B}(H)$. But from the diagrams

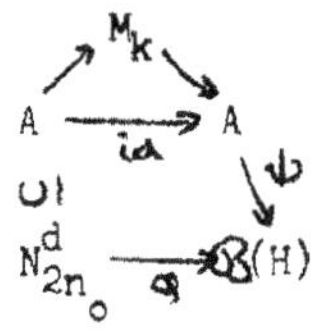

this would imply that φ is nuclear, a contradiction.

Realizing $N_{2n_o}^d$ as an operator system, it will generate a separable C*-algebra. The latter cannot be imbedded in a nuclear C*-algebra. The existence of such a C*-algebra was first proved by Blackadoor [8], employing, however, the continuum hypothesis.

Perhaps, $N_{2n_o}^d$ is a "canonical bad apple" in the sense that

Conjecture: A C*-algebra is non-nuclear if and only if it contains $N_{2n_o}^d$.

The importance of this conjecture is that if it is true we would have:

Consequence of Conjecture: Any C*-subalgebra of a nuclear C*-algebra must again be nuclear.

The Conjecture is rather analogous to the well-known conjecture that a discrete group is non-amenable if and only if it contains the free group on two generators. A C*-algebraic precedent already exists. Letting $\mathfrak{U}$ be a UHF algebra, a modification of Glimm's famous construction [26] shows (see [16,§4.1])

Theorem 10.3: A C*-algebra A is not of type I if and only if there exists a matricial order isomorphism of $\mathfrak{U}$ into A.

Perhaps the most important reason for studying the systems N_{2n}^{d} is that by using them, one might be able to find better intrinsic characterizations for the nuclears and injective. Examining the proof of Theorem 10.1 in [40], we find that the equality of tensor products immediately implies the Riesz decomposition property, and thus is lattice ordering in the dual space. An analogue of this result would give an affirmative answer to a question that has tantalized the author for over five years: is there a matricial analogue of lattice ordering which can be used to characterize the injectives and the nuclears?

11. Appendix of terminology and notation

Real or complex vector spaces V and V_1 are in duality if there is a pairing $(v,f) \to v\cdot f$ from $V \times V_1$ to the scalars, i.e., a bilinear map with $v = 0$ if and only if $v\cdot f = 0$ for all $f \in V_1$, and $f = 0$ if and only if $v\cdot f = 0$ for all $v \in V$. If V and V_1 are in duality, each defines the weak topology on the other. If V, V_1 and W, W_1 are such dual pairs, we let $\mathcal{B}(V,W)$ denote the weakly continuous functions from V to W, and regarding $\mathbb{R}$ (resp., $\mathbb{C}$) as self-dual, we let $V^\delta = \mathcal{B}(V,\mathbb{R})$ (resp., $\mathcal{B}(V,\mathbb{C})$) denote the dual of V. We may identify V_1 with V^δ and V_1^δ with V. We shall use the notation V^d for the algebraic dual of V, and if V is a normed vector space, V^* for the Banach dual.

Given a vector space V, we let $M_n(V)$ denote the linear space of $n\times n$ matrices $v = [v_{ij}]$, $v_{ij} \in V$, and we let $M_n = M_n(\mathbb{C})$. The latter has the canonical basis of matrix units ε_{ij}. If $\varphi: V \to W$ is linear, we define the linear map $\varphi_m : M_m(V) \to M_m(W)$ by $\varphi_m[v_{ij}] = [\varphi(v_{ij})]$. If V and V^δ are in duality, then the same is true for $M_m(V)$ and $M_m(V^\delta)$ under the pairing

$$[v_{ij}][f_{ij}] = \sum f_{ij}(v_{ij})$$

A real ordered vector space is a real vector space V together with a distinguished cone V^+. Dual real ordered vector spaces V and V^δ are dual vector spaces each of which is ordered, such that $V^{\delta +}$ is the dual cone of V^+ and vice versa. Since the dual cone is just the negative polar, this means each cone is weakly closed. In particular, if one starts off with a cone in V, and takes the dual cone in V^δ, one must use the weak closure of V^+ to obtain dual real ordered vector spaces.

A $*$-vector space V is a complex vector space together with a conjugate linear map $v \mapsto v^*$ such that $v^{**} = v$. If $v^* = v$, we say that v is self-adjoint, and we let V_h denote the real linear subspace of such elements. The usual real imaginary part argument shows that $V = V_h + iV_h$. We let $M_n(V)$ have the *-operation $[v_{ij}]^* = [v_{ji}^*]$. Dual *-vector spaces V and V^δ are *-vector spaces with a duality satisfying $v \cdot f^* = (v^* \cdot f)^-$ (complex conjugate). For such spaces, it is immediate that V_h and $(V^\delta)_h$ are in duality. Given two dual *-vector pairs V, V^δ and W, W^δ we define a *-operation on $\mathcal{B}(V,W)$ by $\varphi^*(v) = \varphi(v^*)^*$.

A complex ordered vector space is a *-vector space V, together with a distinguished cone $V^+ \subseteq V_h$. Two such spaces are said to be dual under a given pairing if they are dual as *-vector spaces, and V_h and $(V^\delta)_h$ are dual as real ordered vector spaces. We say that $\varphi \in \mathcal{B}(V,W)$ is positive if $\varphi = \varphi^*$ and $\varphi(V^+) \subseteq W^+$.

We say that a complex vector space V is matrix ordered if

(MI) V is a *-vector space (hence so is $M_n(V)$ for each $n \geq 1$)

(MII) each $M_n(V)$, $n \geq 1$ is ordered

(MIII) if γ is any $m \times n$ matrix of complex numbers,

$$\gamma^* M_m(V)^+ \gamma \subseteq M_n(V)^+$$

(matrix multiplication having the obvious interpretation).

Two such spaces V and V^δ are dual under a given pairing if for each n, $M_n(V)$ and $M_n(V^\delta)$ are dual complex ordered vector spaces. Given two such pairs V, V^δ and W, W^δ, we say that $\varphi \in \mathcal{B}(V,W)$ is completely positive if $(\varphi)_n \geq 0$ for all n. If φ is a linear homeomorphism, and both φ and φ^{-1} are completely positive, we say that φ is a complete or matricial order isomorphism. M_n is itself matrix ordered (one has $M_k(M_n) \cong M_{kn}$), and the natural pairing between M_n and itself is a pairing of matrix ordered spaces (see [12,§4]).

The completely positive maps to or from M_n have an elegant characterization which plays a very important role in the theory. Given dual matrix ordered spaces V and V^δ and elements $v \in M_n(V)$ and $f \in M_n(V^\delta)$, we define $\Theta(v): M_n \to V$ and $\Lambda(f): V \to M_n$ by

$$\Theta(v)(\alpha) = \sum \alpha_{ij} v_{ij}$$
$$\Lambda(f)(v) = [v \cdot f_{ij}].$$

Ordering (M_n, V) and (V, M_n) by the cones of completely positive maps, we have [12, §4]

Lemma 11.1: $\Theta : M_n(V) \to \mathcal{B}(M_n, V)$ and $\Lambda : M_n(V^\delta) \to \mathcal{B}(V, M_n)$ are order isomorphisms.

If V and W are real or complex vector spaces, we let $V \otimes W$ denote their algebraic tensor product. If V and W are $*$-vector spaces, the map $v \otimes w \mapsto v^* \otimes w^*$ determines a $*$-operation on $V \otimes W$. We have that

$$(V \otimes W)_h = V_h \otimes W_h$$

since given $u = \sum v_k \otimes w_k \in (V \otimes W)_h$, adding u to $u^* = u$ and dividing by 2 gives

$$u = \sum \operatorname{Re} v_k \otimes \operatorname{Re} w_k - \sum \operatorname{Im} v_k \otimes \operatorname{Im} w_k.$$

For convenience we shall use the notation

$$V \odot W = (V \otimes W)_h.$$

If V, V^δ and W, W^δ are dual $*$-vector pairs, we let $(V \otimes W)^\delta$ denote the linear functionals on $V \otimes W$ which are continuous in each variable. We have a natural $*$-injection $V^\delta \otimes W^\delta \to (V \otimes W)^\delta$, and $V \otimes W$ has distinct dualities with each of these spaces. If these are dual matrix ordered spaces, it is possible to define various matrix orderings on the tensor products and their duals. However, all we shall need is the corresponding scalar orderings (i.e., $n = 1$) and in fact, we shall restrict our attention to defining orderings on $V \odot W$ and the corresponding duals $(V \odot W)^\delta$, $V^\delta \odot W^\delta$ in §7.

References

1. E. M. Alfsen, Compact Convex Sets and Boundary Integrals, Springer Verlag, Berlin, 1971

2. E. M. Alfsen and F. W. Shultz, State spaces of Jordan algebras, to appear.

3. W. B. Arveson, Subalgebras of C*-algebras. I, Acta Math. 123 (1969), 141-224.

4. W. B. Arveson, Notes on extensions of C*-algebras, Duke Math. J.44(1977), 329-355.

5. W. Bade, The Banach Space C(S), Aarhus University Lecture Notes Series No. 26, 1971.

6. S. Banach, Théorie des Opérations Linéaires, Warsaw, 1932.

7. B. Blackadar, Infinite tensor products of C*-algebras, to appear.

8. B. Blackadar, to appear.

9. M. Choi, A Schwarz inequality for positive linear maps on C*-algebras, Illinois J. Math. 18 (1974), 565-574.

10. M. Choi, Completely positive linear maps on complex matrices, Linear Algebra and Appl. 10 (1975), 285-290.

11. M. Choi, to appear.

12. M. Choi and E. Effros, Injectivity and operator spaces, J. Fnal. Anal. 24 (1974), 156-209.

13. ____________, Nuclear C*-algebras and the approximation property, Amer. J. Math., to appear.

14. ____________, Separable nuclear C*-algebras and injectivity, Duke. Math. J. 43 (1976), 309-322.

15. ____________, Nuclear C*-algebras and injectivity: the general case, Indiana Un. Math. J., 26 (1977), 443-446.

16. ____________, The completely positive lifting problem for C*-algebras, Annals of Math. 104 (1976), 585-609.

17. ____________, Lifting problems and the cohomology of C*-algebras, Can. J. Math., 29 (1977), 1092-1111.

18. G. Choquet, Lectures on Analysis, Benjamin, New York, 1969.

19. A. Connes, Classification of injective factors, Annals of Math. 104 (1976), 73-116.

20. J. Dixmier, Sur les structures Boreliennes du spectra d'une C*-algebre, Inst. Hautes Etudes Sci Publ. Math. 6(1960), 5-11.

21. ________, Les C*-algebres et leurs representations, Gauthier Villars, Paris, 1964.

22. E. Effros, A decomposition theory for representations of C*-algebras, Trans. Amer. Math. Soc. 107 (1963), 83-106.

23. ________, Structure in simplexes, Acta Math. 117 (1967), 103-121.

24. E. Effros and F. Hahn, Locally Compact Transformation Groups and C*-algebras, Memoirs of the Amer. Math. Soc. No. 75, Providence, 1967.

25. E. Effros and C. Lance, Tensor products of operator algebras, Advances in Math. 25,(1977), 1-34.

26. J. Glimm, On a certain class of operator algebras, Trans. Amer. Math. 95 (1960), 318-340.

27. ________, Type I C*-algebras, Annals of Math. 73 (1961), 572-612.

28. D. A. Goodner, Projections in normed linear spaces, Trans. Amer. Math. Soc. 69 (1950), 89-108.

29. F. Greenleaf, Invariant Means on Topological Groups, van Nostrand, New York, 1969.

30. A. Guichardet, Tensor Products of C*-algebras, Aarhus Univ. Lecture Notes Series 12, 1969.

31. J. Hakeda and J. Tomiyama, On some extension property of von Neumann algebras, Tohoku Math. J. 19 (1967), 315-323.

32. Harish-Chandra, Representations of semisimple Lie groups III, Trans. Amer. Math. Soc. 76 (1954), 234-253.

33. R. Kadison, A Representation Theory for Commutative Topological Algebra, Memoirs of the Amer. Math. Soc. No. 7, New York, 1951.

34. ________, Isometries of operator algebras, Annals of Math. 54 (1951), 325-338.

35. ________, A generalized Schwarz inequality and algebraic invariants for operator algebras, Annals of Math. 56 (1952), 494-503.

36. J. Kelley, Banach spaces with the extension property, Trans. Amer. Math. Soc. 72 (1952), 323-326.

37. C. Lance, On nuclear C*-algebras, J. Fnal. Anal. 12 (1973), 157-176.

38. A. Lazar and J. Lindenstrauss, Banach spaces whose duals are L_1 spaces and their representing matrices, Acta Math. 126 (1971), 165-193.

39. R. Lipsman, Group Representations, Lecture Notes in Mathematics No. 388, Springer-Verlag, New York, 1974.

40. I. Namioka and R. Phelps, Tensor products of compact convex sets, Pacific J. Math. 31 (1969), 469-480.

41. L. Pukanszky, Characters of connected Lie groups, to appear.

42. L. Nachbin, A theorem of Hahn-Banach type for linear transformations, Trans. Amer. Math. Soc. 68 (1950), 28-46.

43. S. Sakai, A characterization of type I C*-algebras, Bull Amer. Math. Soc.72 (1966), 508-512.

44. J. Schwartz, Two finite, non-hyperfinite, non-isomorphic factors, Comm. Pure and Appl. Math. 16 (1963), 19-26.

45. W. Stinespring, Positive functions on C*-algebras, Proc. Amer. Math. Soc. 6 (1955), 211-216.

46. M. Stone, Applications of the theory of Boolean rings to general topology, Trans. Amer. Math. Soc. 41 (1937), 375-481.

47. J. Thayer, Exterior orderings and completely positive liftings, to appear.

48. D. Voiculescu, A non-commutative Weyl-von Neumann Theorem, Rev. Roum. Math. Pures et Appl. 21 (1976), 97-113.

49. S. Wasserman, On tensor products of certain group C*-algebras, J. Fnal. Anal. 23 (1976).

50. __________, Injective W*-algebras, Math. Proc. Camb. Phil. Soc. 82 (1977) 39-47.

51. J. Anderson, A C*-algebra A for which Ext A, is not a group, to appear.

52. W. Arveson, Notes on extensions of C*-algebras, to appear.

53. D. Voiculescu, On a theorem of M. D. Choi and E. G. Effros, to appear.

54. R. Loebl, Injective von Neumann algebras, Proc. Amer. Math. Soc. 44 (1974), 46-48.

55. O. Hustad, Intersection properties of balls in complex Banach spaces whose duals are L_1 spaces, Acta Math. 132 (1974), 283-313.

56. A. Lima, Intersection properties of balls and subspaces in Banach spaces, to appear.

59. J. Tomiyama, Tensor products and projections of norm one in von Neumann algebras, Lectore notes for seminar given at Un. of Copenhagen, 1970.

60. J. Glimm, Families of induced representations, Pacific J. Math. 12 (1962), 885-911.

61. T. Huruya, C*-algebras having the property T, Sci. Rep. Nijgata Univ., 1971

CORRESPONDENCES BETWEEN von NEUMANN ALGEBRAS AND DISCRETE AUTOMORPHISM GROUPS

Hisashi Choda
Department of Mathematics
Osaka Kyoiku University
Osaka, Japan.

1. Introduction. F. Murray and J. von Neumann constructed examples of factors by means of a measure space construction in [32]. In 1958, T. Turumaru [47] defined an abstract discrete crossed product of von Neumann algebras and pointed out that the measure space construction of Murray-von Neumann is the crossed product of an abelian von Neumann algebra by an ergodic group of freely acting automorphisms. The discrete crossed products of von Neumann algebras have been used to construct factors and studied for their own interest, from an algebraic point of view by M. Nakamura and Z. Takeda [35] and N. Suzuki [42] (for some additional references, see [20] and [21]).

Recently, the crossed product of a von Neumann algebra by a locally compact group was defined by M. Takesaki [44]. Many interesting results in the theory of continuous crossed products of von Neumann algebras have been obtained by several authors (e.g., [16], [30], [31], [33] and [44], for other references, see [40]).

In this talk, we shall restrict our attention to the discrete crossed product of von Neumann algebras. We consider the relations between von Neumann algebras and automorphism groups. We shall treat only separable Hilbert spaces and discrete countable groups.

2. Discrete crossed products. We begin by stating the definition of a crossed product (following M. Takesaki [44]). Let A be a von Neumann algebra acting on a Hilbert space $\mathcal{H}$, G a group of automorphisms of A and $G\otimes\mathcal{H}$ the Hilbert space of $\mathcal{H}$-valued square summable functions on G. Then the space $G\otimes\mathcal{H}$ is identified with $\mathcal{H}\otimes l^2(G)$ or $\Sigma_{g\varepsilon G}\mathcal{H}\otimes\varepsilon_g$, where $\{\varepsilon_g\}_{g\varepsilon G}$ is an orthonormal basis for the space $l^2(G)$ of square summable functions on G defined by

$$\varepsilon_g(h) = \delta_g^h = \{\begin{smallmatrix} 1 & (g=h) \\ 0 & (g\neq h) \end{smallmatrix} \qquad (g,h\varepsilon G).$$

Define a faithful normal representation π of A on $G\otimes\mathcal{H}$ by

$$(\pi(a)\xi)(h) = h^{-1}(a)\xi(h) \qquad (a\varepsilon A,\ h\varepsilon G,\ \xi\varepsilon G\otimes\mathcal{H})$$

and a unitary representation λ of G on $G\otimes\mathcal{H}$ by

$$(\lambda(g)\xi)(h) = \xi(g^{-1}h) \qquad (g,h\varepsilon G,\ \xi\varepsilon G\otimes\mathcal{H}).$$

Then the pair $\{\pi,\lambda\}$ of representations π of A and λ of G is a covariant representation of $\{A,G\}$;that is,

$$\lambda(g)\pi(a)\lambda(g)^* = \pi(g(a)) \qquad (a\varepsilon A,\ g\varepsilon G).$$

The von Neumann algebra on $G\otimes\mathcal{H}$ generated by $\pi(A)$ and $\lambda(G)$ is called the crossed product of A by G and denoted by $G\otimes A$. For each subgroup K of G, the von Neumann algebra on $G\otimes\mathcal{H}$ generated by $\pi(A)$ and $\lambda(K)$ is denoted by $N(K)$.

Ostensibly, the crossed product $G\otimes A$ depends also the underlying Hilbert space. However, the next proposition assures us that the algebraic structure of the crossed product is independent of the Hilbert space. The proposition is used to prove the uniqueness of the crossed product (within an isomorphism) in several papers (e.g., [38] for II_1-factors, [21] and [44] for general cases).

Proposition 1. Let A_1 and A_2 be von Neumann algebras and σ an isomorphism of A_1 onto A_2. Let G_1 and G_2 be groups of automorphisms of A_1 and A_2 respectively and ϕ an isomorphism of G_1 onto G_2 such that

$$\sigma(g(a)) = \phi(g)(\sigma(a)) \qquad (a\varepsilon A_1,\ g\varepsilon G_1).$$

Then there is an isomorphism α of $G_1\otimes A_1$ onto $G_2\otimes A_2$ such that

$$\alpha(\pi_1(a)) = \pi_2(\sigma(a)) \qquad (a\varepsilon A_1)$$

$$\alpha(\lambda_1(g)) = \lambda_2(\phi(g)) \qquad (g\varepsilon G_1),$$

where $\{\pi_i,\lambda_i\}$ is the covariant representation of $\{A_i,G_i\}$ cited above.

The discrete crossed products of von Neumann algebras are relatively easy to handle, because we have explicit descriptions of elements in the crossed product as Fourier expansions or matrices. We shall describe a matrix representation for each element in the crossed product.

For each $g\varepsilon G$, let $J_g\eta = \eta\otimes\varepsilon_g$ $(\eta\varepsilon\mathcal{H})$. Then J_g is an isometric linear mapping of $\mathcal{H}$ onto the subspace $\mathcal{H}\otimes\varepsilon_g$ of $G\otimes\mathcal{H}$. For each x in the algebra $L(G\otimes\mathcal{H})$ of all operators on $G\otimes\mathcal{H}$, we represent x by a matrix $(x_{g,h})_{g,h\varepsilon G}$, where

$$x_{g,h} = J_g{}^*xJ_h \qquad (g,h\varepsilon G).$$

For each $a\varepsilon A$, we have

$$\pi(a)_{g,h} = \delta_g^h g^{-1}(a) \qquad (g,h\varepsilon G);$$

and, for each $k\varepsilon G$,

$$\lambda(k)_{g,h} = \delta_g^{kh} \qquad (g,h\varepsilon G).$$

Modifying Zeller-Meier [49, Proposition 8.4], we have the following proposition.

Proposition 2. For each subgroup K of G, the von Neumann algebra $N(K)$ is the set of all elements in $L(G\otimes\mathcal{H})$ with the matrix form

$$x_{g,h} = g^{-1}(x(gh^{-1}))$$

for some A-valued function $x(g)$ on G such that $\chi_K(g)x(g) = x(g)$, where χ_K is the characteristic function of K.

An expectation on a von Neumann algebra is a useful tool for investigating the relation between the von Neumann algebra and its subalgebras (cf. [34], [46] and [48]). Let C be a von Neumann algebra and D a von Neumann subalgebra of C. Then a positive linear mapping e of C onto D is called an expectation of C if e satisfies that

$$e(1) = 1$$

and

$$e(cd) = e(c)d \qquad (c\varepsilon C,\ d\varepsilon D).$$

By using Arveson's expectation [3, 6.1.3], we have the following proposition (cf. [9, Proposition 2]).

Proposition 3. For each subgroup K of G, there exists a faithful normal expectation e of $G\otimes A$ onto $N(K)$ such that

$$e(\lambda(g)) = 0 \qquad (g\notin K).$$

In particular, there exists a faithful normal expectation e of $G\otimes A$ onto $\pi(A)$ such that $e(\lambda(g)) = 0$ for all $g \neq 1$.

Conversely, this property characterizes a crossed product [11,

Theorem 4] (also, cf. [15, Proposition 4.1.2]):

Theorem 4. Let M be a von Neumann algebra, A a von Neumann subalgebra of M and G a group of automorphisms of A. Assume that (M, A, G) satisfies the following three conditions;

(1) there exists a unitary representation u_g of G in M such that $g = \mathrm{Ad}\, u_g$ $(g \in G)$, where $\mathrm{Ad}\, u_g(a) = u_g a u_g^*$ $(a \in A)$,

(2) there exists a faithful normal expectation e of M onto A such that $e(u_g) = 0$ for all $g \neq 1$,

(3) M is generated by A and u_G.

Then M is isomorphic to the crossed product $G \otimes A$ by a mapping α such that

$$\alpha(a) = \pi(a) \qquad (a \in A)$$

and

$$\alpha(u_g) = \lambda(g) \qquad (g \in G).$$

If a group G is freely acting on A (cf. §3), then, for each $g \in G$, the dependent element $e(u_g)$ of g is 0. Therefore, we may weaken assumption (2) for a freely acting automorphism group of A in Theorem 4 as follows ([12, Corollary 5]):

(2') there exists a faithful normal expectation of M onto A.

Theorem 4 is fundamental for the sequel. Some II_1-factor versions of this theorem can be found in [20] and [42]. Some generalizations of this theorem for discrete crossed products with factor sets are given in [13] and [41]. A continuous crossed product version is Landstad's characterization of crossed products by locally compact groups [31, Theorem 1].

3. Freely acting automorphisms. The free action of automorphisms of von Neumann algebras played an important role from the beginning in the theory of von Neumann algebras. The notion of free action is used, especially, in connection with crossed products. In this section, we shall consider the free action of automorphisms of von Neumann algebras.

Let A be a von Neumann algebra and α an automorphism of A. Then an element a of A is called a <u>dependent element</u> of α if

$$ab = \alpha(b)a \qquad (b \in A)$$

[14]. Dependent elements have the following properties (cf. [4, §2], [8, Theorem 1] and [14]):

Proposition 5. Let α be an automorphism of a von Neumann algebra A. Then a dependent element a of α has the following properties;

(1) $a^*a = aa^*$

(2) a^*a is a central element of A

(3) $\alpha(a) = a$

and

(4) a^* is a dependent element of α^{-1}.

An automorphism α of a von Neumann algebra A is said to be <u>freely acting</u> on A if there are no nonzero dependent elements of α. In other words, an element $a \in A$ having the property $ab = \alpha(b)a$ for all $b \in A$ is 0 [27]. Suppose that an automorphism α of A is spatial, that is, $\alpha = \mathrm{Ad}\, u$ for some unitary operator u. Then the automorphism α is freely acting on A if and only if $uA' \cap A = \{0\}$ [8, Proposition 1]. Hence, the automorphism α is freely acting on A if and only if the automorphism $\mathrm{Ad}\, u$ of A' is freely acting on A' ([25, Lemma 9], [26, Lemma 1.1]). The following decomposition theorem of automorphisms is well known:

Kallman's Theorem ([27, Theorem 1.11]). Let α be an automorphism of a von Neumann algebra A. Then there exists a unique central projection p in A such that

$\alpha(p) = p$,

α is inner on A_p

and

α is freely acting on $A_{(1-p)}$.

We shall call the projection p in the Kallman's theorem a <u>central projection inducing the inner part</u> of α and denote it by $p(\alpha)$. Then, by Proposition 5, we have the following proposition (cf. [8]):

Proposition 6. Let A be a von Neumann algebra and G a group of automorphisms of A. If an automorphism α of A satisfies $\alpha g = g\alpha$ for all $g \in G$, then $g(p(\alpha)) = p(\alpha)$ for all $g \in G$.

Corollary ([7], [45]). Let G be an outer automorphism group of a von Neumann algebra A. If there exists an automorphism group H of A with the following properties;

(1) $gh = hg \quad (g \in G,\ h \in H)$

and

(2) H is ergodic on $A \cap A'$,

then G is freely acting on A (that is, every g ($\neq 1$) in G is freely acting on A).

In particular, an ergodic and abelian automorphism group of an abelian von Neumann algebra is freely acting.

The notion of free action of an automorphism group is characterized in terms of crossed products by Y. Haga and Z. Takeda [25] and Y. Haga [24]:

Proposition 7. Let G be a group of automorphisms of a von Neumann algebra A. Then the group G is freely acting if and only if

$$G \otimes A \cap \pi(A)' \subset \pi(A).$$

4. Full groups. H. A. Dye introduced the notion of full groups for automorphism groups on abelian von Neumann algebras in [18]. And Y. Haga and Z. Takeda generalized the notion for automorphism groups of general von Neumann algebras in [25].

Let A be a von Neumann algebra and G a group of automorphisms of A. Then we shall denote by $[G]$ the set of all automorphisms α of A having $\sup_{g \in G} p(\alpha^{-1} g) = 1$. We call $[G]$ the <u>full group</u> determined by G. The full group has the following properties (cf. [18] and [25]):

Proposition 8.

(1) $[G]$ is again a group of automorphisms of A.

(2) $[[G]] = [G]$.

(3) The elements α of $[G]$ are precisely those automorphisms of A having a representation

$$\alpha(a) = \Sigma_n\, p_n g_n(vav^*) \qquad (a \in A),$$

where $g_n \in G$, v is a unitary operator in A and $\{p_n\}$ (resp. $\{g_n^{-1}(p_n)\}$) is a family of mutually orthogonal central projections having sum 1.

Suppose that the group G is freely acting on A. For each

$\alpha \varepsilon [G]$, let

$$p_g = g(p(g^{-1}\alpha)) \qquad (g\varepsilon G).$$

Then we have that

$$\alpha(a) = \Sigma_{g\varepsilon G}\, p_g g(vav^*) \qquad (a\varepsilon A)$$

for some unitary operator v in A. We may assume that each $\alpha\varepsilon[G]$ is an inner automorphism of $G \otimes A$ induced by a unitary operator

$$\Sigma_{g\varepsilon G}\, \lambda(g)\pi(p(g^{-1}\alpha)v)$$

for some unitary operator v in A. Conversely an automorphism of this type is in $[G]$. Therefore, we have the following theorem [25, Theorem 1]:

Theorem 9. An automorphism α of A belongs to $[G]$ if and only if α can be extended to an inner automorphism of $G \otimes A$, identifying A with $\pi(A)$.

We shall denote by $[G]_Z$ the set of all automorphisms α of A having the following form

$$\alpha(a) = \Sigma_n\, p_n g_n(a) \qquad (a\varepsilon A),$$

where $g_n \varepsilon G$, and $\{p_n\}$ (resp. $\{g_n^{-1}(p_n)\}$) is a family of mutually orthogonal central projections having sum 1 (cf. [25]). We shall call $[G]_Z$ the <u>Z-full group</u> determined by G.

H. A. Dye introduced equivalent relations among automorphism groups in [18]. Let A_1 and A_2 be von Neumann algebras, and G_1 and G_2 groups of automorphisms of A_1 and A_2 respectively. Then two groups G_1 and G_2 are called <u>weakly equivalent</u> if there exists an isomorphism σ of A_1 onto A_2 such that $[G_1] = [\sigma^{-1}G_2\sigma]$. In the case where $A_1 = A_2$, two groups G_1 and G_2 are called <u>equivalent</u> if $[G_1] = [G_2]$. In terms of crossed products, we shall give a characterization of the equivalence among groups of freely acting automorphisms of an abelian von Neumann algebra; and we shall give a characterization of weak equivalence (cf. [5] and [6]).

Theorem 10. Let A be an abelian von Neumann algebra, and G_1 and G_2 be groups of freely acting automorphisms of A. Then a necessary and sufficient condition that G_1 and G_2 are equivalent is that there exists an isomorphism σ of $G_1 \otimes A$ onto $G_2 \otimes A$ such that

$$\sigma(\pi_1(a)) = \pi_2(a) \qquad (a\varepsilon A).$$

In fact, suppose that $[G_1] = [G_2]$. Then each g_2 in G_2 belongs to $[G_1]$. For each $g_2 \varepsilon G_2$, let

$$v_{g_2} = \Sigma_{g\varepsilon G_1} \lambda_1(g)\pi_1(p(g_2^{-1}g)).$$

Then $g \mapsto v_g$ is a unitary representation of G_2 in $G_1 \otimes A$ and $G_1 \otimes A$ is generated by $\pi_1(A)$ and v_{G_2}. Therefore, by Theorem 4, we have the desired isomorphism of $G_1 \otimes A$ onto $G_2 \otimes A$. The converse is an immediate consequence of Theorem 9.

Theorem 11. Let A_1 and A_2 be abelian von Neumann algebras, and G_1 and G_2 be groups of freely acting automorphisms of A_1 and A_2 respectively. Then a necessary and sufficient condition that G_1 and G_2 are weakly equivalent is that there exists an isomorphism σ of $G_1 \otimes A_1$ onto $G_2 \otimes A_2$ such that $\sigma(\pi_1(A_1)) = \pi_2(A_2)$.

W. Krieger, in [28], gives, in terms of Krieger factors, a similar characterization of weak equivalence. In [29], he also gets a strong version of Theorem 11 for singly generated groups.

5. Correspondences between subgroups and subalgebras in a crossed product. In [35], M. Nakamura and Z. Takeda showed that there is a one-to-one correspondence between the class of subgroups of a group and the class of intermediate subalgebras of the crossed product of a II_1-factor by the group. And Y. Haga and Z. Takeda showed in [25] for finite von Neumann algebras that there is a one-to-one correspondence between the class of full subgroups and the class of intermediate subalgebras of a crossed product, as a generalization of the Dye correspondence [19, Proposition 6.1]. The conservation of types in the Dye-Haga-Takeda correspondence was obtained in [19, Proposition 6.1], [10, Theorem 24] and [23, Theorem 4.10].

In this section, we shall consider a correspondence between subgroups and intermediate subalgebras of a crossed product for a general von Neumann algebra. Detailed accounts of the material we shall discuss in this section and the next section can be found in [9].

We shall need the following lemma, which is a variant of [15, Lemma 1.5.6] and [25, Lemma 5].

Lemma. Let A be a von Neumann algebra, B a von Neumann subalgebra of A with $B' \cap A \subset B$, C a von Neumann subalgebra of A containing B and e an expectation of A onto C. If a unitary operator u in A satisfies the condition $uBu^* = B$, then $e(u)$ has the following properties;

(1) $e(u)$ is a partial isometry,

(2) the initial projection p and the final projection q of $e(u)$ are contained in the center of B, and

(3) $e(u) = up = qu$.

Let A be a von Neumann algebra and G a group of automorphisms of A. We shall call a von Neumann subalgebra of $G \otimes A$ containing $\pi(A)$ an <u>intermediate subalgebra</u> of $G \otimes A$.

Theorem 12. Let G be a group of freely acting automorphisms of a von Neumann algebra A. If there exists a group H of automorphisms of $G \otimes A$ having the following properties;

(1) $h(\pi(A)) = \pi(A) \qquad (h \in H)$,

(2) H is ergodic on $\pi(A \cap A')$

and

(3) $h(\lambda(g)) = \lambda(g) \qquad (g \in G,\ h \in H)$,

then there exists a ono-to-one correspondence between the class of subgroups of G and the class of H-invariant intermediate subalgebras B having a faithful normal expectation of $G \otimes A$ onto B, obtained by associating with each subgroup K the intermediate subalgebra $N(K)$ and, with the intermediate subalgebra B, the subgroup $\{g \in G;\ \lambda(g) \in B\}$ (denoted by $K(B)$).

In fact, by Proposition 3, for each subgroup K of G, $N(K)$ is an H-invariant intermediate subalgebra of $G \otimes A$ having a faithful normal expectation of $G \otimes A$ onto $N(K)$; and we have $K(N(K)) = K$. Conversely, let B be an H-invariant intermediate subalgebra of $G \otimes A$ having a faithful normal expectation e of $G \otimes A$ onto B. For each $g \in G$, by the preceding lemma and freeness of action of G, there exists a central projection p_g in A such that $e(\lambda(g)) = \pi(p_g)\lambda(g)$. On the other hand, by the property that $B' \cap G \otimes A \subset B$, we have that

$$h(e(x)) = e(h(x)) \qquad (x \in G \otimes A,\ h \in H)$$

(cf. [15, Theorem 1.5.5]). Hence, $p_g = 0$ or 1 for each $g \in G$. Then $e(\lambda(g)) = 0$ for all $\lambda(g) \notin B$. Therefore, we have $e(G \otimes A) \subset N(K(B))$.

Corollary. Let A be a factor and G a group of outer automorphisms of A. Then there exists a one-to-one correspondence between the class of subgroups of G and the class of intermediate subalgebras B of $G \otimes A$ having a faithful normal expectation of $G \otimes A$ onto B, by the same association as in Theorem 12.

The Corollary is a generalization of [12, Theorem 5] and [35, Theorem 2] and also has another generalization, which is proved, in a manner similar to that of the proof of Theorem 12.

Theorem 13. Let A be a von Neumann algebra and G a group of (not necessarily freely acting) automorphisms of A. Then there exists a one-to-one correspondence between the class of subgroups of G and the class of intermediate subalgebras B of $G \otimes A$ having a faithful normal expectation e of $G \otimes A$ onto B such that $e(\lambda(g))=0$ for all $\lambda(g) \notin B$, by the same association as in Theorem 12.

6. Galois correspondences. In this section, we shall consider a Galois correspondence. Throughout this section, we shall use a von Neumann algebra A, a group G of freely acting automorphisms of A and the fixed point algebra B of A under G. We shall call a von Neumann subalgebra of A containing B an <u>intermediate subalgebra</u> of A.

M. Nakamura and Z. Takeda introduced a Galois theory for II_1-factors and finite groups of outer automorphisms by observing the close analogy between the theories of classical simple rings and of II_1-factors ([36], [37] and others). M. Henle [26] established, among other results, a Galois correspondence between the class of subgroups of G and a class of intermediate subalgebras of A, under the condition that there exists a family of mutually orthogonal projections p_g $(g \in G)$ in A such that $\Sigma_{g \in G}\, p_g = 1$ and that $g(p_h) = p_{hg^{-1}}$ $(g, h \in G)$. As a consequence of a generalization of the Dye correspondence, Y. Haga and Z. Takeda established a Galois correspondence between the class of Z-full subgroups of Z-full group determined by G and the class of intermediate subalgebras of A, under certain conditions [25].

We shall consider the following conditions;

(A) there exists a faithful normal expectation of B' onto A',

(B) there exists a unitary representation u_g of G such that $g = \mathrm{Ad}\, u_g$ for all $g \in G$,

(C) under the condition (B), there exists an isomorphism θ of B' onto $G \otimes A'$ such that $\theta(A') = \pi(A')$ and that $\theta(u_g) = \lambda(g)$ for all $g \in G$, and

(D) there exists a group H of automorphisms of A such that $gh = hg$ for all $g \in G$ and $h \in H$ and that H is ergodic on the center of A.

The condition (A) is an algebraic property. Under the condition (B), the condition (A) is equivalent to the condition (C) (as follows from Proposition 3 and Theorem 4).

Under the conditions (B) and (C), an automorphism of A leaving B pointwise fixed is in $[G]_Z$ [26, Proposition 2.1]. Then, by Proposition 6, we have the following;

Proposition 14. Under the conditions (B), (C) and (D), if an automorphism α of A satisfies the conditions,

$$h\alpha = \alpha h \qquad (h\varepsilon H)$$

and

$$\alpha(x) = x \qquad (x\varepsilon B),$$

then α is in G.

This proposition is a discrete version of [2, Theorem III.3.3 (i)].

Theorem 15. Suppose the conditions (A) and (D) hold. Then there exists a one-to-one Galois correspondence between the class of subgroups of G and the class of H-invariant intermediate subalgebras C of A having a faithful normal expectation of B' onto C'.

In fact, we may assume that there exists a cyclic and separating vector ξ for A, since the conditions (A) and (D) are algebraic. Considering the canonical unitary implementation u of the automorphism group of A with respect to ξ in [1] (also, cf. [22]), the condition (B) is satisfied by G. Put $H_0 = \{\theta \circ \mathrm{Ad}\, u_h \circ \theta^{-1};\ h\varepsilon H\}$, where θ is the mapping in the condition (C). Then H_0 is a group of automorphisms of $G \otimes A'$ having the properties (1), (2) and (3) in Theorem 12. The theorem follows now from Theorem 12.

Corollary 1. Suppose the condition (D) holds. If the group G is finite, then there exists the same Galois correspondence as in Theorem 15.

Corollary 2. Suppose the condition (A) holds. If the algebra A is a factor, then there exists a one-to-one Galois correspondence between the class of subgroups of G and the class of intermediate subalgebras C of A having a faithful normal expectation of B' onto C'.

Corollary 2 leads us to the following;

Proposition 16. Suppose the conditions (A) and (B) hold. Then there exists a one-to-one Galois correspondence between the class of subgroups of G and the class of intermediate subalgebras C of A having a faithful normal expectation e of B' onto C' such that $e(u_g) = 0$ for all $u_g \notin C'$.

This proposition is a slight generalization of a theorem of Henle [26, Theorem 3.1]. It is valid also for a group G which is not necessarily freely acting, if the following condition (A') is assumed in place of the condition (A):

(A') Under the condition (B), there exists a faithful normal expectation e of B' onto A' such that $e(u_g) = 0$ for all $g \neq 1$.

References

1. H. Araki, Some properties of modular conjugation operator of von Neumann algebras and a non commutative Radon-Nikodym theorem with a chain rule, Pacific J. Math., 50(1974), 309-354.
2. H. Araki, D. Kastler, M. Takesaki and R. Haag, Extention of KMS states and chemical potential, Commun. Math. Phys., 53(1977), 97-134.
3. W. Arveson, Analyticity in operator algebras, Amer. J. Math., 89 (1967), 578-642.
4. H. Behncke, Automorphisms of crossed products, Tōhoku Math. J., 21(1969), 580-600.
5. H. Choda, On the crossed product of abelian von Neumann algebras, I., Proc. Japan Acad., 43(1967), 111-116.
6. ______, On the crossed product of abelian von Neumann algebras, II., Proc. Japan Acad., 43(1967), 198-201.
7. ______, On ergodic and abelian automorphism groups of von Neumann algebras, Proc. Japan Acad., 47(1971), 982-985.
8. ______, On freely acting automorphisms of operator algebras, Kōdai Math. Sem. Rep., 26(1974), 1-21.
9. ______, A Galois correspondence in a von Neumann algebra, Preprint (1976).
10. M. Choda, On types over von Neumann subalgebras and the Dye correspondence, Publ. RIMS, Kyoto Univ., 9(1973), 45-60.
11. ______, Normal expectations and crossed products of von Neumann algebras, Proc. Japan Acad., 50(1974), 738-742.
12. ______, Correspondence between subgroups and subalgebras in the compact crossed product of a von Neumann algebra, Math. Japon., 21(1976), 51-59.
13. ______, Some relations of II_1-factors on free groups, Preprint (1 977).
14. M. Choda, I. Kasahara and R. Nakamoto, Dependent elements of an automorphism of a C*-algebra, Proc. Japan Acad., 48(1972), 561-565.
15. A. Connes, Une classification des facteurs de type III, Ann. Sci.

Ecole Norm. Sup., 6(1973), 133-252.
16. A. Connes and M. Takesaki, The flow of weights on factors of type III, Preprint, UCLA, (1975).
17. J. Dixmier, Les Algebres d'Operateurs dans l'Espace Hilbertien, Gauthier-Villars, Paris, 1957.
18. H. A. Dye, On groups of measure preserving transformations, I., Amer. J. Math., 81(1959), 119-159.
19. _____, On groups of measure preserving transformations, II., Amer. J. Math., 85(1963), 551-576.
20. V. Ya. Golodets, Crossed products of von Neumann algebras, Russian Math. Surveys, 26(1971), 1-50.
21. A. Guichardet, Systèmes dynamiques non commtatifs, Soc. Math. France, 1974.
22. U. Haagerup, The standard form of von Neumann algebras, Math. Scand., 37(1975), 271-283.
23. Y. Haga, On subalgebras of a cross product von Neumann algebra, Tōhoku Math. J., 25(1973), 291-305.
24. _____, Crossed products of von Neumann algebras by compact groups, Tōhoku Math. J., 28(1976), 511-522.
25. Y. Haga and Z. Takeda, Correspondence between subgroups and subalgebras in a cross product von Neumann algebra, Tōhoku Math. J., 24(1972), 167-190.
26. M. Henle, Galois theory of W*-algebras, Preprint.
27. R. R. Kallman, A generalization of free action, Duke Math. J., 36(1969), 781-789.
28. W. Krieger, On constructing non-*isomorphic hyperfinite factors of type III, J. Functional Analysis, 6(1970), 97-109.
29. _____, On ergodic flows and the isomorphism of factors, Preprint.
30. A. Kishimoto, Remarks on compact automorphism groups of a certain von Neumann algebra, Preprint (1976).
31. M. Landstad, Duality theory of covariant systems, Trondheim Univ., Preprint (1974).
32. F. J. Murray and J. von Neumann, On rings of operators, Ann. Math., 37(1936), 116-229.
33. Y. Nakagami, Essential spectrum $\Gamma(\beta)$ of a dual action on a von Neumann algebra, Preprint (1976).
34. M. Nakamura and T. Turumaru, Expectations in an operator algebra, Tōhoku Math. J., 6(1954), 182-188.
35. M. Nakamura and Z. Takeda, On some elementary properties of the crossed products of von Neumann algebras, Proc. Japan Acad., 34(1958), 489-494.
36. _____ and _____, A Galois theory for finite factors, Proc. Japan Acad., 36(1960), 258-260.
37. _____ and _____, On the fundamental theorem of the Galois theory for finite factors, Proc. Japan Acad., 36(1960), 313-318.
38. T. Saito, The direct product and the crossed product of rings of operators, Tōhoku Math. J., 11(1959), 299-304.
39. I. M. Singer, Automorphisms of finite factors, Amer. J. Math., 77(1955), 117-133.
40. S. Strătilă, D. Voiculescu and L. Zsidó, On crossed products, Rev. Roum Math. Pures et Appl., 21(1976), 1411-14449 and 22 (1977), 83-117.
41. C. E. Sutherland, Cohomology and extensions of von Neumann algebras, II., Univ. of Oslo, Preprint (1976).
42. N. Suzuki, Crossed products of rings of operators, Tōhoku Math. J., 11(1959), 113-124.
43. H. Takai, On a Fourier expansion in continuous crossed products, Publ. RIMS, Kyoto Univ., 11(1976), 849-880.
44. M. Takesaki, Duality for crossed products and the structure of von Neumann algebras of type III, Acta Math., 131(1973), 249-310.

45. P. K. Tam, On ergodic abelian M-group, Proc. Japan Acad., 47(1971), 456-457.
46. J. Tomiyama, Tensor products and projections of norm one in von Neumann algebras, Lecture Note at Univ. of Copenhagen, (1970).
47. T. Turumaru, Crossed product of operator algebra, Tōhoku Math. J., 10(1958), 355-365.
48. H. Umegaki, Conditional expectation in an operator algebra, III., Kōdai Math. Sem. Rep., 11(1959), 51-64.
49. G. Zeller-Meier, Produits croisés d'une C*-algèbre par un groupe d'automorphisms, J. de Math. Pure et Appl., 47(1968), 101-239.

The construction and decomposition of quantum fields using operator theory, probability and fiber bundles

J. Glimm
Rockefeller University
New York 10021, N.Y.

Supported in part by the National Science Foundation under grant PHY76-17191

The construction of quantum fields in two and three space time dimensions is now completed. The frontier questions concern four dimensions as well as the details of the structure of the two and three dimensional models. For these latter problems, there has been considerable progress. A major conclusion which has emerged from this work is that the formal constructions of theoretical physics are mathematically consistent, at least as far as the program has progressed. (In two and three dimensions, one has verified the axioms, constructed particles with a nontrivial S-matrix and possible bound states, etc. In four dimensions the mass and wave function renormalization problems have been solved, but not the charge renormalization.) The mathematics necessary to perform this task is more singular than previous theories, but because of its internal symmetry and rich structure, parts of it may find a place in the general mathematical culture.

The construction is easiest to describe in the Euclidean region ($t \to it$) as we now explain. For d space time dimensions, we begin with the Schwartz space $\mathcal{S}(R^d)$ of rapidly decreasing smooth test functions and its dual $\mathcal{S}'(R^d)$, of tempered distributions. The elements

$$\varphi(\vec{x},t) \in \mathcal{S}'(R^d)$$

may be regarded as generalized functions

$$\varphi(\vec{x},\cdot) : t \to \varphi(\vec{x},t)$$

defined on the real line and taking values in $\mathcal{S}'(R^{d-1})$.

We now introduce an integration theory over $\mathcal{S}'(R^d)$; for comparison with the case of Wiener measure, $\varphi(\vec{x},\cdot)$ is the path, $\mathcal{S}'(R^d)$ is the path space, and (up to unresolved technical problems) $\mathcal{S}'(R^{d-1})$ should be thought of as the state space for the stochastic process we are about to construct. Explicitly, we postulate the existence of a positive Borel measure $d\varphi$

on $\mathcal{S}'(R^d)$ with the following properties:

1. Normalization: $\int d\phi = 1$

2. Symmetry: $\int F(\phi)d\phi = \int F(U\phi)d\phi$

where U is a Euclidean transformation of R^d (rotation, translation or reflection)

$$(U\phi)(\vec{x},t) \equiv \phi(U^{-1}(\vec{x},t)) .$$

3. Regularity.

4. Physical positivity.

5. Ergodicity: The unitary group acts ergodically on the measure space $\mathcal{S}'(R^d)$, $d\phi$.

The regularity condition is technical and may be formulated in a variety of ways, but always includes a condition such as

$$e^{<\phi f>} \in L_1(d\phi) \text{ , } \forall f \in \mathcal{S}(R^d)$$

which guarantees existence of moments. The physical positivity is crucial. To explain it, we introduce the time zero reflection operator Θ:

$$(\Theta\phi)(x,t) = \phi(x,-t)$$

$$\Theta F(\phi) = F(\Theta\phi) .$$

Let $\mathcal{E} = L_2(\mathcal{S}',d\phi)$ and let $\mathcal{E}^+$ be the subspace of $\mathcal{E}$ generated by all functions of the positive time fields,

$$\left\{\phi(x,t): \ t > 0 \right\} .$$

Then physical positivity is the condition

$$0 \leq \int \Theta F(\phi)^{-} F(\phi) d\phi, \quad \forall F \in \mathcal{E}^{+} .$$

The construction of the quantum mechanical Hilbert space $\mathcal{H}$ and the quantum field Φ, (which is an unbounded operator valued distribution acting on $\mathcal{H}$) is modeled on the C^*-algebra construction of a representation from a C^*-algebra state. Explicitly,

$$\mathcal{N} = \{F \in \mathcal{E}^{+} : \int \Theta F^{-} F d\phi = 0$$

and $\mathcal{H}$ is the completion of $\mathcal{E}^{+}/\mathcal{N}$ in the topology defined by the inner product

$$\mathcal{E}^{+} \ni F, G \rightarrow \langle F, G \rangle_{\mathcal{H}} \equiv \int \Theta F^{-} G d\phi .$$

In this construction, the unitary group $U(t)$ of time translations on $\mathcal{E} = L_2(d\phi)$ maps into a self adjoint contraction semigroup e^{-tH} on $\mathcal{H}$ with generator H, and H is the Hamiltonian of the field theory. Ignoring some technical complications, we also define

$$\Phi(x,t) = e^{itH} \phi(x,0) e^{-itH}$$

and state the main reconstruction theorem.

Theorem 1. Assuming 1)-5), Φ satisfies all Wightman axioms.

A general introductory reference, including a self contained proof of this theorem and references may be found in [6].

A simple example of a measure $d\phi$ which satisfies the Osterwalder-Schrader axioms 1)-5) is the Gaussian Ornstein-Uhlenbeck measure with covariance operator

$$C = (-\Delta + m^2)^{-1} .$$

The resulting field Φ is the free field. Nontrivial examples require more work, and we only write down the suggestive formal expression

$$(1) \quad d\phi \sim e^{-\int [\nabla\phi^2 + \frac{1}{2}m^2\phi + \mathcal{P}(\phi)]\, dx} \prod_x d\phi(x) .$$

Theorem 2. For $d = 2$ and P semibounded and for $d = 3$ and $P = \phi^4$, the expression $d\phi$ above can be given a mathematical meaning and 1)-4) are valid. Analogous results for the $d = 2$ Yukawa field theory are also valid.

References: $P(\phi)_2$ [9,2] ; Y_2 [17, 12, 18] ; ϕ_3^4 [14, 18].

The main idea is to introduce cutoffs ($R^d \to T^d$, the d-torus; $T^d \to T^d \cap \varepsilon Z^d$, a finite lattice) and then prove uniform estimates which allow removal of the cutoff in all renormalized expressions. In this way infinite subtractions are given a rigorous mathematical meaning.

Theorem 3. $d\phi$ as above. Condition 5) is valid in some cases, not in others.

References: [19, 9, 10, 11, 3].

The breakdown of 5) is associated with the occurrence of phase transitions, and is resolved on an abstract level by appeal to the theory of direct integral decompositions of von Neumann algebras [1, 4]. It was proved by Araki (under general hypothesis, verified in most of the above cases) that the algebra of observables associated with $d\phi$ via Theorem 1 is type I with an abelian commutant $\mathcal{R}'$. In the resulting decomposition

$$(2) \quad \mathcal{H} = \int_{\xi \varepsilon \Xi} \oplus\, d\mathcal{H}_\xi$$

one has explicit information (and still more explicit conjectures) about Ξ which come from methods of statistical mechanics. An appealing intuitive picture comes from the kinetic energy term $(\nabla\phi)^2$ in (1) and the entropy resulting from the potential energy

$$V(\phi) = \tfrac{1}{2}m^2\phi+P(\phi) .$$

Assuming V to have the simple form in Fig. 1 below

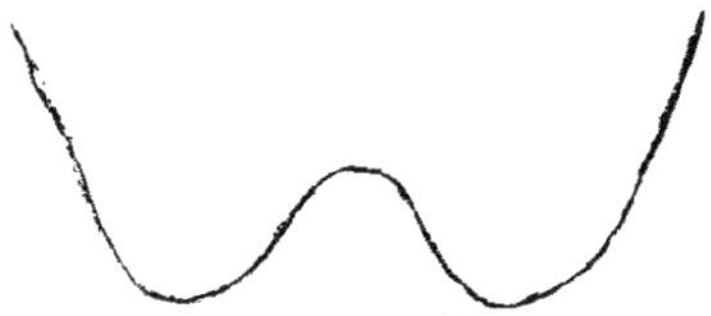

Fig. 1. $V(\phi) = \phi^4-\phi^2$,

V is minimized when $\phi = \pm 1$. The choice of sign ($\phi(x)\simeq\pm1$, for each x) is governed by entropy, and there are only two possibilities.

If kinetic energy is dominant, then $(\nabla\phi)^2 \simeq 0$ and $\phi \simeq$ constant, i.e., the choice $\phi(x)\simeq\pm1$ is predominantly independent of x, and $d\phi$ in the decomposition (2) is a convex sum: $d\phi = \alpha_+ d\phi_+ + \alpha_- d\phi_-$. A set of measure one for $d\phi_+$ has $\phi\simeq-1$ on small islands, and $\phi \cong +1$ elsewhere, as illustrated in Fig. 2.

Fig. 2. A typical configuration in the measure $d\phi_+$. $\pm$ indicates points where $\phi\simeq\pm 1$.

If entropy is dominant, the density of $\pm$ islands has increased to the point where there is no coherent choice of a + sign (no connected "ocean" complementary to the islands). In this case Ξ in (2) consists of a single point, and condition 5) is valid.

With no effort, one can imagine P's having 1, 2, 3, ... , (deg P)/2 global minima, resulting in a generalization of the above picture. As the coefficients of P are varied, generically Ξ has a single point, corresponding to a unique global minimum of P, while phase transitions correspond to a shift in the absolute minimum from one local minimum to another, and occur for P's having two or more global minima. However quantum corrections complicate this picture, so that exact locations of the phase transitions are given only by a divergent infinite series. A partial justification of this picture is contained in [11].

As a concluding section, I would like to explain two facts of life which have motivated recent work directed towards the $d = 4$ problem [15, 5, 16, 13, 7, 8]. First, renormalization in four dimensions is qualitatively more difficult than for $d = 1, 2, 3$. The infinite part of the renormalization is only defined implicitly for $d = 4$. Consequently one must return to first principles, which state that the $\xi \to 0$ limit is equivalent to the study of some critical point of a lattice (i.e. ultraviolet cutoff) field theory.

A critical point, in our above picture of phase transitions, occurs when two global minima coincide (e.g. $V(\phi) \equiv \phi^4$).

This notion agrees with the definition of critical point in the sense of Morse theory. For any V, the mass m, (up to quantum corrections, as always) is given by

$$m = V''(\phi)$$

evaluated at $\phi = \phi_{min}$, i.e. at the global minimum. Hence we see that $m = 0$ at the critical point, and the approach to a critical point can be characterized by $m \to 0$. Herein lies the advantage of the critical point for the mass renormalization of the ε (= lattice spacing) $\to 0$ limit. If the coefficients of V are varied to approach the critical point (for some fixed lattice spacing ε) and simultaneously all distances are scaled, so that

$$\varepsilon \to \varepsilon' \equiv s\varepsilon \qquad m \to m' \equiv s^{-1}m$$

then we may choose $s = \text{const.}\ m \to 0$, so that after scaling

$$\varepsilon' \to 0, \quad m' = \text{const.}$$

In other words the critical point of the lattice theory provides an exact solution of the mass renormalization problem of the corresponding continuum theory, cf. [15] . For a ϕ^4 theory, the field strength renormalization is accomplished in [5], but the problem of charge renormalization remains open. Uniform estimates to control the $\varepsilon \to 0$ limit are also needed, and a number of partial results are known, as described in [6] . The remaining estimates are a consequence of a conjectured correlation inequality: $\Gamma^{(6)} \leq 0$.

The second fact of life is that the interactions considered

above, e.g. $\phi^4 \oplus$ Yukawa, are at best an incomplete description of strongly interacting particles, and appear on physical grounds to be inappropriate for processes to which weak interactions contribute. As a physically more fundamental type of interaction, the coupled gauge-fermion (quark) interaction has been proposed. Guided by the experience with the ϕ^4 interaction, we expect that a qualitative study of phase transitions and critical points for the lattice guage-quark system will be a part of solving the renormalization problem for these interactions.

A guage field A is by definition a connection form in a fiber bundle. As such it is a Lie algebra valued 1-form. Its covariant differential $F = DA$ is the curvature tensor, and it also takes values in a Lie algebra. The Yang Mills field equations

$$DF = D^2A = 0, \quad \delta F = 0 \tag{2}$$

are the Euler equations which result from minimizing the action

$$\mathcal{Q} = \int_{R^4} \mathrm{Tr} F \cdot F \, dx .$$

Thus this structure is determined by the choice of the Lie group G. $G = T^1$, the one-torus, leads to Maxwell's equations and the photon field. In the gauge-quark system, G is chosen compact, semisimple and nonabelian, which implies that the equations (2) are nonlinear. Finally (Euclidean) quantization amounts to the construction of a measure dA on the space of all possible A's (not restricted by the condition $\delta DA = 0$) and generalizing (1) in the sense that

$$dA \sim e^{-\mathcal{Q}} \prod_{x \in R^4} dA(x).$$

BIBLIOGRAPHY

1. J. Fröhlich, Schwinger Functions and their generating functionals II, Adv. Math, to appear.

2. J. Fröhlich, B. Simon, Pure states for general $P(\varphi)_2$ theories: construction, regularity and variational equality. Ann. Math to appear.

3. J. Fröhlich, B. Simon and T. Spencer, Infrared bounds, phase transitions and continuous symmetry breaking. Commun. Math. Phys 50, 79-95 (1976)

4. J. Glimm and A. Jaffe, Two and three body equations in quantum field theory. Commun Math Phys. 44, 293-275 (1975)

5.-----, Critical exponents and elementary particles. Commun. Math. Phys. 52, 203-210 (1976).

6.-----, In: 1976 Cargèse lectures

7.-----, Quark trapping for lattice U(1) gauge theories. Physics letters 66B, 67-69 (1977).

8.-----, Instantons in a U(1) lattice gauge theory. Preprint.

9. J. Glimm, A. Jaffe and T. Spencer, In: Constructive quantum field theory, G. Velo and A. Wightman (eds.) Springer lecture notes in physics vol. 25, Springer-Verlag, 1973.

10.-----, Phase transitions for φ^4 quantum fields, Commun. Math. Phys. 45, 203-216 (1975).

11.-----, A convergent expansion about mean field theory I,II. Ann. Phys. 101, 610-669 (1976).

12. O. McBryan, Volume dependence of Schwinger functions in the Yukawa quantum field theory. Commun. Math. Phys. 45, 279-294 (1975).

13. O. McBryan and T. Spencer, On the decay of correlations in symmetric ferromagnets. Commun. Math. Phys., 53, 299-302 (1977).

14. Park, Uniform bounds on the pressure of the φ^4_3 field model. J. Math. Phys. 17. 1073-1075. (1976).

15. J. Rosen, Mass renormalization of the φ^4 Euclidean field theories. Adv. Math., to appear.

16. R. Schrader, A possible constructive approach to φ^4_4 III. 50, 97-102 (1976).

17. E. Seiler, Schwinger functions for the Yukawa model in two dimensions with space time cutoff. Commun. Math. Phys. 42, 163-182 (1975).

18. E. Seiler and B. Simon, Nelson's symmetry and all that in Yukawa and ϕ_3^4 field theories, Ann. Phys. 97, 470–518 (1976).

19. B. Simon and R. Griffiths, The φ_2^4 field theory as a classical Ising model. Commun. Math. Phys. 33, 145–164 (1973).

On KMS states of a C* dynamical system*

Huzihiro ARAKI

Research Institute for Mathematical Sciences
Kyoto University, Kyoto 606, Japan

Abstract Some topics related to KMS states of a C* dynamical system are reviewed. In particular, conditions equivalent to the KMS condition are discussed. For the most part, the C* dynamical system is taken to be a UHF algebra.

1. KMS states of a C* dynamical system

A C* dynamical system is a pair consisting of a C*-algebra $\underline{A}$ and a one-parameter group α_t, $t \in R$ of automorphisms of $\underline{A}$ such that for each $a \in A$, $\alpha_t a$ is continuous in $t \in R$. In the following, we assume that $\underline{A}$ is separable and unital ($1 \in A$).

A state φ of $\underline{A}$ is called an (α_t, β)-KMS state (or simply a KMS state) if it satisfies the following KMS-condition (at the inverse temperature β), where β is a real number:

For every pair of elements a and b of $\underline{A}$, there exists a function $F_{a,b}(z)$ of a (complex) variable z defined on the closed set

$$I_\beta \equiv \{z : \text{Im } z \in [0,\beta] \text{ or } [\beta,0]\} \tag{1.1}$$

such that (1) $F_{a,b}(z)$ is bounded and continuous on I_β, (2) $F_{a,b}(z)$ is holomorphic in the interior of I_β (this condition being empty for $\beta = 0$) and (3) the boundary values are given by

$$F_{a,b}(t) = \varphi(a\alpha_t b), \; F_{a,b}(t+i\beta) = \varphi((\alpha_t b)a). \tag{1.2}$$

This condition was first introduced in quantum statistical mechanics by Kubo [28], Martin and Schwinger [31] as a condition satisfied by thermodynamic Green's functions and was first formulated in C*-algebra language by Haag, Hugenholtz and Winnink [22], who called it the Kubo-Martin-Schwinger boundary condition. It is found by Takesaki [47] that the modular automorphisms in the Tomita-Takesaki theory satisfy (and for a given state are uniquely characterized by) the KMS condition at $\beta = -1$. (In theory of C*-algebras, $\beta = -1$ unless otherwise stated.)

In statistical mechanics, α_t describes the time development of the system and the KMS condition is considered to be a condition characterizing the state of the system at thermal equilibrium (with the absolute temperature $T = (k\beta)^{-1}$ where k is the Boltzmann constant). A justification for this from the stability point of view was given recently by Haag, Kastler and Trych-Pohlmeyer [23]. (Also see [17])

Remark 1. If $\underline{A}$ is the algebra of all $n \times n$ matrices ($n < \infty$), then there exists $H \in A$ such that $\alpha_t(a) = e^{itH} a e^{-itH}$, $a \in A$. Then there is a unique (α_t, β)-KMS state and it is given by the following Gibbs state:

$$\varphi_\beta^G(a) = \mathrm{Tr}(e^{-\beta H} a)/\mathrm{Tr}(e^{-\beta H}). \tag{1.3}$$

Remark 2. The KMS condition at $\beta = 0$ is nothing but the condition for a trace state. If α_t is trivial, then the KMS conditions at all β are mutually equivalent.

Remark 3. An (α_t, β)-KMS state φ is α_t-stationary (i.e. $\varphi \alpha_t = \varphi$, $t \in R$). Therefore, α_t has a unique extension $\tilde{\alpha}_t$ to *-automorphisms of the weak closure $\pi_\varphi(A)''$ in the cyclic representation associated with φ. The cyclic vector Ω_φ associated with φ is (cyclic and) separating for $\pi_\varphi(A)''$ and the associated modular automorphisms coincide with $\tilde{\alpha}_{-\beta t}$. This enables one to use properties of modular operators in the analysis of KMS states.

Remark 4. The set of (α_t, β)-KMS states is compact, convex and a Choquet simplex [38]. An (α_t, β)-KMS state is extremal if and only if it is primary (i.e. the center of $\pi_\varphi(A)''$ is trivial). The integral decomposition of a KMS state into extremal KMS states coincides with the central decomposition. In particular $\tilde{\alpha}_t$ of Remark 3 acts trivially on the center of $\pi_\varphi(A)''$. These results appeared in early studies of KMS states [49], [2], [30], [47], [3].

Remark 5. If φ is an (α_t, β)-KMS state, then the function

$$(\alpha_{t_1}(a_1) \dots \alpha_{t_n}(a_n))$$

possesses an analyticity property in $(t_1, \dots, t_n)$ similar to the KMS condition [1].

2. Alternative forms of the KMS condition

Let $\tilde{f} \in \mathcal{D}$,

(2.1) $$f_\gamma(t) = (2\pi)^{-1} \int e^{-ipt+\gamma p} \tilde{f}(p)\, dp,$$

(2.2) $$b(f_\gamma) = \int f_\gamma(t) \alpha_t(b)\, dt.$$

Then the KMS condition at β is equivalent to

(2.3) $$\varphi(ab(f_0)) = \varphi(b(f_\beta)a)$$

for every $a, b \in A$ and $\tilde{f} \in \mathcal{D}$. It is also equivalent to the same condition with $a = b^*$.

For a given stationary state φ, there exist probability measures μ_a and ν_a for each $\underline{a}$ satisfying

(2.4) $$\mu_a(\tilde{f}) = \varphi(a^* a(f_0)), \quad \nu_a(\tilde{f}) = \varphi(a(f_0)a^*).$$

Then the KMS condition at β is equivalent to the two conditions: (1) φ is stationary and (2) for every $a \in A$, μ_a and ν_a are equivalent measures with

(2.5) $$d\mu_a / d\nu_a = e^{\beta p}.$$

The above two alternative forms, though useful, are minor modifications of the KMS condition and all involve the action of α_t for all t. The following condition recently introduced by Sewell [43] involves only the generator δ of α_t and hence is of great significance. By definition, if $\alpha_t a$ has a derivative at $t = 0$, then $\underline{a}$ is in the domain $D(\delta)$ of the generator δ of α_t, and

(2.6) $\delta a = (d/dt)\alpha_t a|_{t=0}$.

$D(\delta)$ is a dense * subalgebra of $\underline{A}$.

Theorem 1. A state φ is an (α_t,β)-KMS state if and only if it satisfies the following Sewell condition for the generator δ of α_t:

Sewell condition: For every $a \in D(\delta)$

(2.7) $\varphi(\delta a) = 0$

(2.8) $-i\beta\varphi(a^*\delta a) \geq S(\varphi(aa^*);\ \varphi(a^*a))$

where

(2.9) $$S(v;u) \begin{cases} = u \log (u/v) & \text{if } u > 0,\ v > 0, \\ = 0 & \text{if } u = 0, \\ = +\infty & \text{if } u > 0,\ v = 0. \end{cases}$$

(Sewell writes the left hand side of (2.8) as

$$(i\beta/2)(\varphi((\delta a^*)a) - \varphi(a^*\delta a)).$$

Note that $\varphi((\delta a^*)a) + \varphi(a^*\delta a) = \varphi(\delta(a^*a)) = 0$.)

We shall briefly indicate the proof for the equivalence of the KMS and Sewell conditions: First note that the condition $\varphi\circ\alpha_t = \varphi$ and $\varphi\circ\delta = 0$ are equivalent as $D(\delta)$ is a dense α_t-invariant subset of $\mathfrak{A}$ and $(d/dt)\varphi\circ\alpha_t = \varphi\circ\delta\circ\alpha_t$. Assuming the KMS condition in the form of (2.5) for φ first, we obtain

$$S(\varphi(aa^*);\varphi(a^*a)) = S(\nu_a(1);\mu_a(1)) = S(\mu_a(k^{-1});\mu_a(1))$$

where $k(p) = \exp \beta p$. By the joint convexity and homogeneity (of degree 1) of the function $S(v;u)$, we obtain

$$S(\Sigma\lambda_j v_j;\ \Sigma\lambda_j u_j) \leq \Sigma\lambda_j S(v_j;\ u_j)$$

for $\lambda_j \geq 0$. By the lower semicontinuity of $S(v;u)$, we obtain

$$S(\mu(f_2);\ \mu(f_1)) \leq \mu(S(f_2;\ f_1))$$

for any positive measure μ and posotive functions f_1, f_2. In particular

$$S(\mu_a(k^{-1});\ \mu_a(1)) \leqq \mu_a(S(k^{-1};1)) = \mu_a(\log k)$$

$$= -i\beta\varphi(a^*\delta a),$$

which proves (2.8). The converse proof is given in [43].

In statistical mechanics, the derivation δ is given on a *-subalgebra A_0 of $\underline{A}$ and sometimes it is not known whether corresponding α_t exists and is unique. The Sewell condition makes sense even for such a situation. (See §6 for a further elaboration on this point.

Another characterization of KMS states given recently by Fannes and Verbeure [50] is the following two conditions

(1) Stationarity : $\varphi\cdot\alpha_t = \varphi$.

(2) Bogoliubov-type inequality :

$$\mu_a(g) \leqq \{\varphi(aa^*) - \varphi(a^*a)\}/\log\{\varphi(aa^*)/\varphi(a^*a)\}$$

for $g(p) = \{1-e^{-\beta p}\}/\beta p$ and $a \in A$ where μ_a is the measure given by (2.4). (The spectral condition in [50] is unnecessary, as the Sewell condition follows immediately from these two conditions.) The Sewell condition was motivated by this work.

3. Ground states

A state φ is called a <u>ground state</u> if

(3.1) $$\varphi(a(f_0)^*a(f_0)) = 0$$

for all $a \in A$ and $\tilde{f} \in \mathcal{D}$ with supp $\tilde{f} < 0$. We shall call a state φ a <u>ceiling state</u> if

(3.2) $$\varphi(a(f_0)^*a(f_0)) = 0$$

for all $a \in A$ and $\tilde{f} \in \mathcal{D}$ with supp $\tilde{f} > 0$.

Both ground and ceiling states are stationary. Let $U_\varphi(t)$ be the one-parameter group of unitaries satisfying

$$U_\varphi(t)\pi_\varphi(a)\Omega_\varphi = \pi_\varphi(\alpha_t a)\Omega_\varphi$$

for a stationary state φ. Then a necessary and sufficient condition for a stationary state to be a ground (ceiling) state is that the self-adjoint generator of $U_\varphi(t)$ is positive (negative).

Ground and ceiling states can be viewed as KMS states with $\beta = +\infty$ and $-\infty$, respectively, as can be seen from the following theorem. ([33], [32])

Theorem 2. <u>If β_ν is a net of real numbers converging to β and φ_ν is a net of (α_t, β_ν)-KMS states weakly converging to φ, then φ is an (α_t, β)-KMS state if β is finite, a ground state if $\beta = +\infty$ and a ceiling state if $\beta = -\infty$</u>

We shall briefly indicate the proof : If $|\beta| < \infty$, f_{β_ν} tends to f_β in L_1-norm and hence $b(f_{\beta_\nu})$ tends to $b(f_\beta)$ strongly in $\mathfrak{A}$. Hence the characterization (2.3) of the KMS condition implies the conclusion of Theorem. If $\beta = +\infty$ and if $\operatorname{supp} \tilde{f} < 0$ for $\tilde{f} \in \mathcal{D}$, then f_{β_ν} tends to 0 in L_1-norm and hence $b(f_{\beta_\nu})$ tends strongly in $\mathfrak{A}$. Again the characterization (2.3) implies (3.1) and hence the conclusion of Theorem. The case of $\beta = -\infty$ is similar.

4. Local structure

Let L be an atomic complete Boolean lattice. Let $N(I)$ denote the cardinality of the set of atoms below $I \in L$. Let L_0 be the set of all I with $N(I) < \infty$. The smallest and largest elements will be denoted by $\emptyset$ and 1_L. In statistical mechanics, the atoms of L will be points of a lattice.

The local structure of $\underline{A}$ is the system of C^* subalgebras $A(I)$, $I \in L$ satisfying the following conditions: If $I = \bigvee_\alpha I_\alpha$, $A(I)$ is generated by $A(I_\alpha)$. If $I = \bigwedge_\alpha I_\alpha$, then $A(I) = \bigcap_\alpha A(I_\alpha)$. $A(\emptyset) = \mathbb{C}1$ and $A(1_L) = A$. If $I_1^c > I_2$, then $A(I_1)$ and $A(I_2)$ commute elementwise where c denotes the orthocomplement.

In the following we assume that $A(I)$, for an atom $I \in L$, is a finite dimensional factor and the number of atoms is denumerably infinite. Hence $\underline{A}$ is a UHF algebra.

Let δ be the generator of α_t and suppose that each $A(I)$, for any atom $I \in L$ (and hence for any $I \in L_0$), is contained in $D(\delta)$. We shall say in this case that δ is 'L-normal'. If each element of $A(I)$, for any atom I (and hence for any $I \in L_0$), is an analytic element of δ in addition, then we shall call δ to be 'L-analytic'. For any given δ_t, for a UHF algebra, there exists an L such that δ

is L-analytic [39].

If $A(\Lambda) \subset D(\delta)$ for $\Lambda \in L_0$, there exists [39] $H(\Lambda) = H(\Lambda)^* \in A(\Lambda)$ such that

(4.2) $$\delta a = i[H(\Lambda), a], \quad a \in A(\Lambda).$$

Such an $H(\Lambda)$ is determined by δ up to addition of elements in $A(\Lambda^c)$

Let ω be a product state of A (namely $\omega(aa') = \omega(a)\omega(a')$ for $a \in A(\Lambda)$, $a' \in A(\Lambda^c)$ and for any $\Lambda \in L$). Let p_Λ^ω denote the conditiona expectation defined for all $a \in A$ by

$$p_\Lambda^\omega(a) \in A(\Lambda^c),$$

(4.3) $$\omega(ab) = \omega(p_\Lambda(a)b), \quad b \in A(\Lambda^c)$$

(If τ denotes the unique trace state of $\underline{A}$, p_Λ^τ is called the partial trace.)

By replacing $H(\Lambda)$ by $H(\Lambda) - p_\Lambda^\omega(H(\Lambda))$, it is always possible to choose $H(\Lambda)$ of (4.2) satisfying

(4.4) $$p_\Lambda^\omega(H(\Lambda)) = 0.$$

The conditions (4.2) and (4.4) uniquely determine $H(\Lambda)$.

Let

(4.5) $$U(\Lambda) = p_{\Lambda^c}^\omega(H(\Lambda)), \; W(\Lambda) = H(\Lambda) - U(\Lambda),$$

(4.6) $$\Phi(\Lambda) = U(\Lambda) - \sum_{\Lambda_1 \lneqq \Lambda} \Phi(\Lambda_1).$$

The function Φ is called the <u>potential</u>. From the definition, it follows that

(4.7) $$\Phi(\emptyset) = 0, \; p_I^\omega(\Phi(\Lambda)) = 0 \quad \text{if} \quad I \wedge \Lambda \neq \emptyset.$$

(4.8) $$U(\Lambda) = \sum_{\Lambda_1 \leqq \Lambda} \Phi(\Lambda_1), \; H(\Lambda) = U(\Lambda) + W(\Lambda)$$

(4.9) $$W(\Lambda) = \lim_{\Lambda' \uparrow} W_{\Lambda'}(\Lambda)$$

where the limit converges in norm and

(4.10) $$W_{\Lambda'}(\Lambda) = \sum\{\Phi(\Lambda_1); \Lambda \wedge \Lambda_1 \neq \emptyset, \; \Lambda^c \wedge \Lambda_1 \neq \emptyset, \; \Lambda_1 \leqq \Lambda'\}.$$

Let us denote by Φ^{δ} the Φ constructed above from δ.

Conversely, let Φ be a function defined on L_0 with the value $\Phi(\Lambda) \in A(\Lambda)$. Suppose (4.7) is satisfied and (4.9) is convergent. Then we define $H(\Lambda)$ by (4.8) and define

$$(4.11) \qquad \delta^{\Phi} a = i[H(\Lambda), a], \quad a \in A(\Lambda).$$

This defines, in a consistent way, a derivation δ^{Φ} with domain

$$(4.12) \qquad A_0 \equiv \bigcup_{\Lambda} \{A(\Lambda);\ \Lambda \in L_0\}.$$

Furthermore $\Phi^{\delta^{\Phi}} = \Phi$ and $\delta^{\Phi^{\delta}} = \delta|_{A_0}$.

We shall briefly sketch proof of some assertions in the above discussion. First assume that $H(\Lambda)$ is given and satisfies (4.2). The consistency of (4.2) requires $H(\Lambda')-H(\Lambda) \in A(\Lambda^c)$ if $\Lambda' \supset \Lambda$. The assumption (4.4) implies the vanishing of $p^{\omega}_{\Lambda_1}(H(\Lambda))$ for any $\Lambda_1 \supset \Lambda$. We define $p^{\omega}_{\emptyset}$ as the identity map and hence $H(\emptyset) = 0$, which implies $\Phi(\emptyset) = 0$. By the consistency condition for $H(\Lambda)$, we have $H(\Lambda)-H(\Lambda_1) \in A(\Lambda_1^c)$ for $\Lambda_1 \leqq \Lambda$ and hence

$$p^{\omega}_{\Lambda_1^c}(H(\Lambda)) - U(\Lambda_1) = p^{\omega}_{\Lambda_1^c}\{H(\Lambda) - H(\Lambda_1)\}$$

$$= \omega(H(\Lambda) - H(\Lambda_1)) = 0.$$

Therefore we obtain the following formula for any $\Lambda_1 \subset \Lambda$:

$$(4.13) \qquad p^{\omega}_{\Lambda\setminus\Lambda_1}(U(\Lambda)) = p^{\omega}_{\Lambda_1^c}(H(\Lambda)) = U(\Lambda_1).$$

To prove the second equation in (4.7), we use an induction on Λ. Let $I \wedge \Lambda \equiv J \neq \emptyset$. Then by the inductive assumption and (4.6),

$$p^{\omega}_{I}(\Phi(\Lambda)) = p^{\omega}_{I\setminus J}\{p^{\omega}_{J}(U(\Lambda)) - \sum_{\Lambda_1 \leqq \Lambda\setminus J} \Phi(\Lambda_1)\}$$

which vanishes by (4.13) for $\Lambda\setminus\Lambda_1 = J$ and (4.6) with Λ replaced by $\Lambda\setminus J$. Hence we have (4.7). The equation (4.8) is an immediate consequence of definitions (4.5) and (4.6).

For $\Lambda' \geqq \Lambda$, (4.13) and (4.4) imply

$$U(\Lambda'\setminus\Lambda) = p^{\omega}_{(\Lambda'\setminus\Lambda)^c}(H(\Lambda'))$$

$$= P^{\omega}_{(\Lambda'\setminus\Lambda)^c}(H(\Lambda') - H(\Lambda))$$

$$= P^{\omega}_{(\Lambda')^c}P^{\omega}_{\Lambda}(H(\Lambda') - H(\Lambda))$$

$$= P^{\omega}_{(\Lambda')^c}(H(\Lambda') - H(\Lambda))$$

where the last equality is due to the consistency condition: $H(\Lambda') - H(\Lambda) \in A(\Lambda^c)$. By (4.10), (4.8), (4.5) and the above computation, we obtain

$$W_{\Lambda'}(\Lambda) = U(\Lambda') - U(\Lambda'\setminus\Lambda) - U(\Lambda) \tag{4.14}$$

$$= P^{\omega}_{(\Lambda')^c}(H(\Lambda')) - P^{\omega}_{(\Lambda')^c}(H(\Lambda') - H(\Lambda)) - U(\Lambda)$$

$$= P^{\omega}_{(\Lambda')^c}(H(\Lambda)) - U(\Lambda).$$

Since $P^{\omega}_{(\Lambda')^c}(a)$ tends to $\underline{a}$ in norm as Λ' tends to 1_L for $a \in A$, we obtain (4.9) by the above computation and (4.5).

For the converse, assume that $\Phi(\Lambda)$ satisfying (4.7) is given and (4.9) is convergent in norm. Then (4.8) defines $H(\Lambda)$. Then (4.6), (4.5) and (4.4) are an immediate consequence of (4.7). Since

$$H(\Lambda) = \lim_{\Lambda''} \{U(\Lambda) + W_{\Lambda''}(\Lambda)\}$$

$$= \lim_{\Lambda''} \sum\{\Phi(\Lambda_1);\ \Lambda_1 \leqq \Lambda'',\ \Lambda_1 \wedge \Lambda \neq \emptyset\},$$

we obtain, for $\Lambda' \geqq \Lambda$,

$$H(\Lambda') - H(\Lambda) = \lim_{\Lambda''} \sum\{\Phi(\Lambda_1);\ \Lambda_1 \leqq \Lambda''\setminus\Lambda,\ \Lambda_1 \wedge \Lambda' \neq \emptyset\}$$

$$\in A(\Lambda^c),$$

which shows the consistency of definition (4.2) of a derivation δ. This completes the sketch of proof.

For a given Φ, the following question is important: Does there exist an α_t whose generator δ coincides with δ^{Φ} on A_0, and whethe such α_t is unique?

In the following two cases, both questions have an affirmative answer and the closure of δ^{Φ} is the generator of the unique α_t, namel A_0 is the core of the generator of α_t.

(1) There exists an increasing sequence $\Lambda_n \nearrow 1_L$ such that $\sup_n \|W(\Lambda_n)\| < \infty$. [26] (This condition applies to one-dimensional lattices in statistical mechanics.)

(2) The following norm is finite for some $s > 0$ [46], [36], [37],

$$(4.15) \qquad \|\Phi\|_s \equiv \sup_x \{ \sum_{\Lambda \geq x} \|\Phi(\Lambda)\| e^{s(N(\Lambda)-1)} \}$$

where x runs over atoms of L.

In the second case, $a \in A_0$ is analytic for δ^Φ (i.e. $\sum |t|^n \|(\delta^\Phi)^n a\| / n! < \infty$ for $|t| < (2\|\Phi\|_s)^{-1} s$).

Remark 1. For a lattice system of spin 1/2, ω is taken to be the trace state τ. Then $\Phi(\Lambda)$ satisfying (4.4) is a sum of terms which are products of Pauli spin matrices, one from every lattice point in Λ. For lattice gas models, ω is taken to be the no-particle state (the state with all spin down).

Remark 2. The mathematical structure discussed in this section is adapted to lattice systems. For continuous systems, see [19].

5. Existence and uniqueness of KMS states

A one-parameter group of automorphisms α_t of a C*-algebra $\underline{A}$ is said to be approximately inner if it is a limit of inner automorphisms:

$$\alpha_t a = \lim_\nu e^{itH_\nu} a e^{-itH_\nu}$$

for some $H_\nu \in A$. For a UHF algebra $\underline{A}$, it is known [33] that α_t is approximately inner if A_0 is the core of the generator δ of α_t. In particular, this holds in the two cases of the preceding sections.

For an approximately inner α_t, there exists at least one (α_t, β)-KMS state for each real β; hence a ground state and a ceiling state [33]. For the case (1) of the preceding section, an (α_t, β)-KMS state is unique for each β [8], [27], [41]. In some examples of quantum statistical mechanics, existence of more than one (α_t, β)-KMS states is known.

It is an open question [33] whether all α_t of a UHF algebra $\underline{A}$ are approximately inner. For this problem as well as for the problem of the preceding section, the study of necessary and sufficient condition for the closure of a derivation such as δ^Φ to be a generator [14],

[15], [16], [18], [20], [21], [40], seems to be important.

6. Gibbs condition and LTS condition

A positive linear functional φ is called separating if Ω_φ is separating for $\pi_\varphi(A)''$ — a condition stronger than faithfulness. For separating φ and $h = h^* \in A$, the perturbed functional φ^h is defined by the vector

$$\Omega_\varphi(h) = \exp\{(\pi_\varphi(h) + \log \Delta_\varphi)/2\}\Omega_\varphi$$

where Δ_φ is the modular operator for Ω_φ. A state φ satisfies the Gibbs condition at β if φ is separating and for every $\Lambda \in L_0$

(6.1) $$\varphi^{\beta W(\Lambda)} = \varphi^G_{\Lambda,\beta} \otimes \varphi'_\Lambda$$

where $\varphi^G_{\Lambda,\beta}$ is the Gibbs state (1.3) of $A(\Lambda)$ with $H = U(\Lambda)$ and φ'_Λ is some positive linear functional on $A(\Lambda^c)$. The equation (6.1) is equivalent to

(6.2) $$\varphi^{\beta H(\Lambda)} = \tau_\Lambda \otimes \varphi''_\Lambda$$

where τ_Λ is the unique trace state of $A(\Lambda)$.

Theorem 3. [6] An (α_t, β)-KMS state satisfies the Gibbs condition at β. If A_0 is the core of the generator of α_t, then any state satisfying the Gibbs condition at β is an (α_t, β)-KMS state.

The density matrix $\rho^\varphi_\Lambda \in A(\Lambda)$ of a state φ is defined by

(6.3) $$\varphi(a) = \tau(\rho^\varphi_\Lambda a), \quad a \in A(\Lambda)$$

where τ is the trace state of A. (This differs in normalization from the conventional definition in terms of the trace of matrices.) The entropy $S_\Lambda(\varphi)$ in Λ (as a closed system) is defined by

(6.4) $$S_\Lambda(\varphi) = -\varphi(\log \rho^\varphi_\Lambda).$$

The entropy $\tilde{S}_\Lambda(\varphi)$ in Λ (as an open system) is defined by

(6.5) $$\tilde{S}_\Lambda(\varphi) = \lim_{\Lambda'\uparrow} \{S_{\Lambda'}(\varphi) - S_{\Lambda'\setminus\Lambda}(\varphi)\},$$

where the expression in parentheses is monotone decreasing. It can also be written as

$$(6.6) \qquad \tilde{S}_{\Lambda}(\varphi) = -S(\tau_{\Lambda} \otimes \varphi_{\Lambda^c} / \varphi)$$

where φ_{Λ^c} is the restriction of φ to $A(\Lambda^c)$ and $S(\psi/\varphi)$ is the relative entropy [9], [10]. The free energy (as open system) is defined by

$$(6.7) \qquad \tilde{F}_{\Lambda,\beta}(\varphi) = \varphi(H(\Lambda)) - \beta^{-1}\tilde{S}_{\Lambda}(\varphi).$$

A state φ satisfies <u>the local thermodynamic stability (LTS) condition at</u> β if

$$(6.8) \qquad \tilde{F}_{\Lambda,\beta}(\varphi) \leq \tilde{F}_{\Lambda,\beta}(\psi)$$

for every $\Lambda \in L_0$ and every ψ satisfying $\psi_{\Lambda^c} = \varphi_{\Lambda^c}$.

Theorem 4. <u>If a state satisfies the Gibbs conditon at</u> β, <u>it satisfies the LTS condition at</u> β [12]. <u>If a state satisfies the LTS condition at</u> β, <u>it satisfies the Sewell condition for</u> δ^{Φ}. [43]

Thus, if A_0 is the core of the generator of α_t, then KMS, Gibbs, LTS and the Sewell conditions are all equivalent.

In a model of statistical mechanics, Φ is given and one has to find out whether α_t exists and is unique. However, Gibbs, LTS and Sewell conditions make sense without α_t. For example, the following easy argument establishes the existence of a state satisfying the Sewell condition for δ^{Φ}, independent of questions about α_t: Let

$$(6.9) \qquad \varphi_{\Lambda,\beta}(a) = \tau(e^{-\beta H(\Lambda)t}a)/\tau(e^{-\beta H(\Lambda)t}).$$

Since $\varphi_{\Lambda,\beta}$ satisfies the KMS condition for $\alpha_t^{\Lambda}(a) = e^{iH(\Lambda)t}ae^{-iH(\Lambda)t}$, it satisfies conditions (2.7) and (2.8) for $a \in A(\Lambda)$. Hence any accumulation point of $\varphi_{\Lambda,\beta}$ as $\Lambda \nearrow 1_L$, which exists by compactness, satisfies the Sewell condition for δ^{Φ}.

A variance of the above argument also proves that any accumulation point of the Gibbs states

$$(6.10) \qquad \varphi^{G}_{\Lambda,\beta}(a) = \tau(e^{-\beta U(\Lambda)t}a)/\tau(e^{-\beta U(\Lambda)t})$$

satisfies the Sewell condition provided that (4.9) converges in norm.

Remark. If $\underline{A}$ has a commutative C^* subalgebra C such that $C \wedge A(\Lambda)$ is maximal abelian in $A(\Lambda)$ and $H(\Lambda) \in C$ for all $\Lambda \in L_0$, we call α_t classical. Let p denote the conditional expectation defined by $p(a) \in C$, $\tau(ab) = \tau(p(a)b)$ for all $b \in C$. A state φ is called classical if $\varphi \circ p = \varphi$. The Gibbs condition for a classical α_t is equivalent to two conditions: (1) φ is classical, (2) The Gibbs condition holds for the restriction of φ to C. The second condition is exactly the same as the DLR-equations. [5]

An interesting problem arises for a case which is intermediate between classical and quantum system. [51]

7. Symmetry group

Let the additive group Z^ν act on L as automorphisms (denoted by $\Lambda \in L \to \Lambda + n \in L$ for $n \in Z^\nu$) such that the action on atoms is free and transitive. ($\Lambda \in L$ is then identified with a subset of Z^ν.) A potential Φ is said to be translationally invariant if there exists a representation of the group Z^ν by automorphisms τ_n, $n \in Z^\nu$, of A satisfying

(7.1) $$\tau_n A(\Lambda) = A(\Lambda + n), \quad \tau_n \Phi(\Lambda) = \Phi(\Lambda + n).$$

Assume that

(7.2) $$\|\Phi\|_{\Lambda_1} \equiv \sum\{N(\Lambda)^{-1} \|\Phi(\Lambda)\| \,;\, \Lambda \wedge \Lambda_1 \neq \emptyset\} < \infty.$$

for $\Lambda_1 \in L_0$. We set $|||\Phi||| = \|\Phi\|_x$ for an atom x, which is independent of the atom x. Then $\|\Phi\|_\Lambda$ and $\|U(\Lambda)\|$ are bounded by $N(\Lambda)|||\Phi|||$.

The limit

(7.3) $$P(\Phi) = \lim N(\Lambda)^{-1} \log \tau(e^{-U(\Lambda)}),$$

as Λ tends to Z^ν, with the surface-to-volume ratio tending to zero (denoted by $\Lambda \nearrow Z^\nu$), exists and defines a continuous convex function on the Banach space B of translationally invariant potentials Φ with the norm $|||\Phi|||$. It is called the presure. A continuous linear functional α on B is called the tangent to P at $\Phi \in B$ if for all Ψ

(7.4) $$P(\Phi + \Psi) \geqq P(\Phi) + \alpha(\Psi).$$

For such α, there exists a unique state φ on A such that φ is translationally invariant ($\varphi \cdot \tau_n = \varphi$) and $\varphi(a) = -\alpha(\Phi_a^\Lambda)$ for $a \in L_0$, where Λ can be any elements of L_0 satisfying $a \in A(\Lambda)$, $\Phi_a^\Lambda(\Lambda + n) = \tau_n a$ and $\Phi_a^\Lambda(\Lambda_1) = 0$ otherwise. We call such φ a tangent state at Φ.

For a translatioanlly invariant state φ, the limits

$$s(\varphi) = \lim N(\Lambda)^{-1} S_\Lambda(\varphi),$$

$$e_\Phi(\varphi) = \lim N(\Lambda)^{-1} \varphi(U(\Lambda))$$

as $\Lambda \nearrow Z^\nu$ exist and are called mean entropy and mean energy. The function P satisfies

$$P(\beta\Phi) = \sup(s(\varphi) - \beta e_\Phi(\varphi))$$

where φ runs over all translationally invariant states of $\underline{A}$.

Theorem 5 [37]. <u>A translationally invariant state</u> φ <u>of</u> A <u>is a tangent state at</u> $\beta\Phi$ <u>if and only if</u>

$$s(\varphi) - \beta e_\Phi(\varphi) = \sup_\psi (s(\psi) - \beta e_\Phi(\psi)). \tag{7.5}$$

This is called the <u>variational principle</u>.

Theorem 6. <u>Assume that</u> Φ <u>is translationally invariant and (4.9) is convergent in addition to (7.2)</u>.

(1) <u>If a translationally invariant state</u> φ <u>satisfies (7.5), then it satisfies the Sewell condition</u>. (cf. [29].)

(2) <u>Any translationally invariant state</u> φ <u>satisfying the Gibbs condition satisfies (7.5)</u>. [7]

(3) <u>Any translationally invariant state</u> φ <u>satisfying the LTS condition satisfies (7.5)</u>. [12]

The proof of this Theorem is standard once we establish the following estimates from the translational invariance of $\Phi(\Lambda)$ (or of $H(\Lambda)$).

$$\|U(\Lambda)\| \leq \|H(\Lambda)\| \leq N(\Lambda)\|\Phi\|, \tag{7.6}$$

$$\lim_\Lambda \|W(\Lambda)\| / N(\Lambda) = 0, \tag{7.7}$$

where the limit of Λ tending to 1_L is the Van Hove limit and $\|\Phi\|$ is a norm on potentials defined by

$$(7.8) \qquad \|\Phi\| = \sup\{\|H(n)\| \; ; \; n = \text{atom}\},$$

which is the same as $H(n)$ for any atom n for a translationally invariant Φ. (Note that potentials Φ for which (4.7) holds and (4.9) converges, or equivalently consistent families $\{H(\Lambda)\}$ satisfying (4.4) form a Banach space relative to $\|\Phi\|$.)

The estimate (7.6) is a simple consequence of (4.5), $\|p_I^{\omega}\| = 1$ for any I and the following formula:

$$(7.9) \qquad H(\Lambda) = H(n_1) + \sum_{j=2}^{N(\Lambda)} p^{\omega}_{(n_1 \vee \ldots \vee n_{j-1})}(H(n_j))$$

where the n_j, $j=1,\ldots,N(\Lambda)$ are distinct atoms contained in Λ. The formula (7.9) follows from the following formulas for $\Lambda' \geqq \Lambda$ which follows from (4.14), (4.8) and (4.10):

$$p^{\omega}_{(\Lambda')^c}(H(\Lambda)) = \sum\{\Phi(\Lambda_1) ; \; \Lambda_1 \leqq \Lambda', \; \Lambda_1 \wedge \Lambda \neq \emptyset\},$$

$$p^{\omega}_{(\Lambda')^c}\{p^{\omega}_{(n_1 \vee \ldots \vee n_{j-1})}(H(n_j))\}$$

$$= \sum\{\Phi(\Lambda_1) ; \; \Lambda_1 \leqq \Lambda' \setminus \{n_1 \vee \ldots \vee n_{j-1}\}, \; n_j \leqq \Lambda_1\}.$$

In this derivation of (7.6), we only used the finiteness of $\|\Phi\|$.

To derive (7.7) for the case where the surface to volume ratio for Λ tends to 0, we divide Λ into a surface layer S which consists of atoms within a fixed distance from some atom on the surface of Λ and the rest $\Lambda \setminus S = K$. By translational invariance, there exist a $\delta > 0$ for any given $\varepsilon > 0$ such that $\|W(n) - W_{\Lambda'}(n)\| < \varepsilon$ if the distance of the surface of Λ' from n is greater than δ. We now use the following identity which is easily established by first considering the same equation with W replaced by $W_{\Lambda''}$ for $\Lambda'' \geqq \Lambda$:

$$(7.10) \qquad W(\Lambda) = (W(n_1) - W_{\Lambda}(n_1)) + \sum_{j=2}^{N(\Lambda)} p^{\omega}_{(n_1 \vee \ldots \vee n_{j-1})}(W(n_j) - W_{\Lambda}(n_j)).$$

For those n_j which are in K, we can use the estimate $\|W(n_j) - W_{\Lambda}(n_j)\| < \varepsilon$ if the thickness of S is greater than δ. For those n_j which are in S, we use the estimate

$$\|W(n)\| = \|H(n) - p_n H(n)\| \leqq 2\|\Phi\|.$$

Then we obtain

$$\|W(\Lambda)\| \leq 2N(S)\|\Phi\| + \varepsilon N(\Lambda).$$

In the limit of $N(S)/N(\Lambda)$ tending to 0, we have

$$\overline{\lim}\, \|W(\Lambda)\| \leq \varepsilon.$$

Since ε is arbitrary, we obtain (7.7). A slight modification of the argument gives (7.7) for the Van Hove limit.

The rest of the proof of Theorem 6 is in the quoted references.

By this Theorem, in cases (1) and (2) of Section 4, the KMS, Gibbs, LTS and Sewell conditions are all equivalent to the variational principle for translationally invariant states.

Here again, the existence of translationally invariant states satisfying the Sewell condition can be shown easily, even if α_t is not known to exist, by considering an accumulation point of the translationally averaged states

$$N(\Lambda)^{-1} \sum_{n\in\Lambda} \varphi^{G}_{\Lambda,\beta}\cdot\tau_n \quad (\text{or} \quad N(\Lambda)^{-1} \sum_{n\in\Lambda} \varphi_{\Lambda,\beta}\,\tau_n)$$

in the limit $\Lambda \nearrow Z^{\nu}$ and by proving that it is translationally invariant and satisfies the Sewell condition due to the convexity of the function S, again provided that (4.9) converges in norm.

8. Discussions

We have discussed several conditions which are closely related. The missing relations seem to be a direct derivation of the Gibbs condition from the LTS and Sewell conditions, of the LTS condition from the Sewell condition, of the Gibbs and LTS conditions from the variational principle and of the variational principle from the Sewell condition, although these relations are already established under various special circumstances through indirect routes. The direct derivation seems to be important in those cases where the existence of α_t is more difficult to obtain. In such a case, a key question is probably whether a state satisfying the Sewell condition is separating and whether the generator of the modular automorphisms coincide with δ^{Φ} for a separating state satisfying the Sewell condition.

There seems to be some pathological behavior (from the viewpoint of statistical mechanics) for potentials restricted only by the requirement $|||\Phi||| < \infty$. [25] Whether the convergence of (4.9) together with

$|||\Phi||| \leq \infty$ still admits such pathological behavior seems to be an interesting question.

There are some results on the extension of KMS states on the G-fixed-point algebra A^G to a state of $\underline{A}$ (theory of chemical potential where G is a compact group of automorphisms of $\underline{A}$ (called gauge group [11], [13]. They will be discussed in the article of D. Kastler.

Takesaki [48] proved that the associated cyclic representations of two states φ and ψ are disjoint if they satisfy the KMS condition at different values of the inverse temperature (φ for $\beta = \beta_1$, ψ for $\beta = \beta_2$ and $\beta_1 \neq \beta_2$) and if the associated representation of one of the states is of type III.

There are some results related to types of von Neumann algebras associated with equilibrium states. Let G be a group of automorphisms of $\underline{A}$ and φ be a primary G-invariant state such that φ is separatin for the weak closure of the associated cyclic representation π_φ of $\underline{A}$ and the cyclic vector Ω_φ associated with φ is the only vector, up to a scalar multiple, invariant under the unitary representation of G canonically defined on the cyclic space associated with φ. Then either φ is a trace state or $\pi_\varphi(A)''$ is a factor of type III. ([24], [44].) The following result directly related to the KMS condition gives a further information ([45], [4].) Let G be an asymptotically abelian group of automorphisms of $\underline{A}$ commuting with α_t. If a G-invariant state φ satisfies the KMS condition relative to α_t, then the S-set of Connes is given by the spectrum of the unitary representation canonically implementing α_t, i.e. $S(\pi_\varphi(A)'') = \mathrm{Spec}\, U_{-\beta t}$ with $U_t\pi_\varphi(a)\Omega_\varphi = \pi_\varphi(\alpha_t a)\Omega_\varphi$ for $a \in A$. If φ is weakly clustering with respect to α_t in addition, either φ is a character or $\pi_\varphi(\mathcal{A})''$ is of type III_1. If a type I_2 factor is in $\mathcal{A}$ in the above situation, $\pi_\varphi(\mathcal{A})''$ satisfi also the property L'_λ for $0 \leq \lambda \leq 1/2$. [52]

References

[1] H. Araki, Publ. RIMS, Kyoto Univ., A4 (1968), 361-371.
[2] H. Araki and H. Miyata, Publ. RIMS, Kyoto Univ., A4 (1968), 373-385
[3] H. Araki, Comm. Math. Phys., 14 (1969), 120-157.
[4] H. Araki, Comm. Math. Phys., 28 (1972), 267-277.
[5] H. Araki and P. D. F. Ion, Comm. Math. Phys., 35 (1974), 1-12.
[6] H. Araki, C*-algebras and their applications to statistical mechani and quantum field theory, ed. D. Kastler. (North Holland Publ. Co, 1976), 64-100.
[7] H. Araki, Comm. Math. Phys., 38 (1974), 1-10.
[8] H. Araki, Comm. Math. Phys., 44 (1975), 1-7.
[9] H. Araki, Publ. RIMS, Kyoto Univ., 11 (1975-76), 809-833.

[10] H. Araki, Publ. RIMS, Kyoto Univ., 13 (1977), No. 1.
[11] H. Araki, and A. Kishimoto, Comm. Math. Phys., 52 (1977), 211-232.
[12] H. Araki and G. L. Sewell, Comm. Math. Phys., 52 (1977), 103-109.
[13] H. Araki, D. Kastler, M. Takesaki and R. Haag, Comm. Math. Phys., 53 (1977), 97-134.
[14] O. Bratteli and D. W. Robinson, Comm. Math. Phys., 42 (1975), 253-268; ibid. 46 (1976), 11-30.
[15] O. Bratteli, Self-adjointness of unbounded derivations on C*-algebras, Marseille Preprint.
[16] O. Bratteli and D. W. Robinson, Comm. Math. Phys., 46 (1976), 31-35.
[17] O. Bratteli and D. Kastler, Comm. Math. Phys., 46 (1976), 37-42.
[18] O. Bratteli and D. W. Robinson, Unbounded derivations of von Neumann algebras, Ann. Inst. H. Poincaré, to appear.
[19] O. Bratteli and D. W. Robinson, Comm. Math. Phys., 50 (1976), 133-156.
[20] O. Bratteli, R. H. Herman and D. W. Robinson, Quasianalytic vectors and derivations of operator algebras, ZIF Preprint.
[21] O. Bratteli and U. Haagerup, Unbounded derivations and invariant states.
[22] R. Haag, N. M. Hugenholtz and M. Winnink, Comm. Math. Phys., 16 (1967), 81-104.
[23] R. Haag, D. Kastler and E. B. Trych-Pohlmeyer, Comm. Math. Phys., 38 (1974), 173-193.
[24] N. M. Hugenholtz, Comm. Math. Phys., 6 (1967), 189-193.
[25] R. B. Israel, Comm. Math. Phys., 43 (1975), 59-68.
[26] A. Kishimoto, Comm. Math. Phys., 47 (1976), 25-32.
[27] A. Kishimoto, Comm. Math. Phys., 47 (1976), 167-170.
[28] R. Kubo, J. Phys. Soc. Japan, 12 (1957), 570-586.
[29] O. E. Lanford and D. W. Robinson, 9 (1968), 327-338.
[30] O. E. Lanford, Systèmes a un nombre infini de degrés de Liberté. (CNRS, Paris, 1970), 146-154.
[31] P. C. Martin and J. Schwinger, Phys. Rev., 115 (1959), 1342-1373.
[32] G. K. Pedersen, Maximal Temperature?, Private circulation.
[33] R. T. Powers and S. Sakai, Comm. Math. Phys., 39 (1975), 273-288.
[34] R. T. Powers and S. Sakai, J. Functional Analysis, 19 (1975), 81-95.
[35] D. W. Robinson, Comm. Math. Phys., 6 (1967), 151-160.
[36] D. W. Robinson, Comm. Math. Phys., 7 (1968), 337-348.
[37] D. Ruelle, Statistical mechanics : rigorous results. (Benjamin, New York, 1969).
[38] D. Ruelle, Cargèse lecture in physics, vol. 4, ed. D. Kastler. (Gordon and Breach Sci. Publ., 1970), 169-194.
[39] S. Sakai, Amer. J. Math., 98 (1976), 427-440.
[40] S. Sakai, Comm. Math. Soc., 43 (1975), 39-40.
[41] S. Sakai, J. Functional Analysis, 21 (1976), 203-208.
[42] S. Sakai, Tohoku Math. J., 28 (1976), 583-590.
[43] G. L. Sewell, KMS conditions and local thermodynamical stability of quantum lattice systems II, to appear in Comm. Math. Phys.
[44] E. Størmer, Comm. Math. Phys., 6 (1967), 194-204.
[45] E. Størmer, Comm. Math. Phys., 28 (1972), 279-294; ibid. 38 (1974), 341-343.
[46] R. F. Streater, Comm. Math. Phys., 6 (1967), 233-247.
[47] M. Takesaki, Tomita's theory of modular Hilbert algebras and its applications. (Lecture Notes in Math. 128, Springer Verlag, Berlin, 1970).
[48] M. Takesaki, Comm. Math. Phys., 17 (1970), 33-41.
[49] M. Winnink, Cargèse lecture in physics, vol. 4 ed. D. Kastler. (Gordon and Breach Sci. Publ., 1970), 235-255.
[50] M. Fannes and A. Verbeure, Correlation inequalities and equilibrium states. (Univ. Leuven preprint.)

[51] A. Kishimoto, Equilibrium states of a semi-quantum lattice system, to appear in Rep. Math. Phys.
[52] D. Testard, Asymptotic ratio set of von Neumann algebras generated by temperature states in statistical mechanics, to appear in Rep. Math. Phys.

RECENT DEVELOPMENTS IN THE THEORY OF UNBOUNDED DERIVATIONS IN C*-ALGEBRAS

Shôichirô Sakai

§1. Introduction. In this talk, I would like to give a brief survey of recent developments in the theory of unbounded derivations in C*-algebras and to discuss some related problems. Because of the diversity of developments, I cannot cover the subject completely within limited time. Many interesting topics are missing from my talk. Nevertheless, I hope, I may expose the scope of the developments.

Let $\mathfrak{A}$ be a C*-algebra. A linear mapping δ in $\mathfrak{A}$ is said to be a *-derivation in $\mathfrak{A}$ if it satisfies the following conditions:

(1) The domain $\mathcal{D}(\delta)$ of δ is a dense *-subalgebra of $\mathfrak{A}$,

(2) $\delta(ab) = \delta(a)b + a\,\delta(b) \qquad (a,b \in \mathcal{D}(\delta))$.

(3) $\delta(a^*) = \delta(a)^* \qquad (a \in \mathcal{D}(\delta))$.

If $\mathcal{D}(\delta) = \mathfrak{A}$, then δ is closed, so that it is bounded [64]. On the other hand, if δ is bounded, then it extends uniquely to a bounded *-derivation on $\mathfrak{A}$; therefore the study of everywhere defined *-derivations is equivalent to the study of bounded *-derivations.

The study of bounded derivations on a C*-algebra is making great strides and is one of the most active branches in the theory of operator algebras. Many capable researchers are contributing to the construction of a beautiful theory of bounded derivations.

If I try to touch on the subject, I will drown in the richness of material. Since the main theme of my talk is the survey of unbounded derivations, I will restrict my discussion of bounded derivations to the mention of two recent beautiful results. One is the solution of the lifting problem for bounded derivations on separable C*-algebras by G. Pedersen [57] and the other is the characterization of separable C*-algebras with the property that every bounded derivation is inner, by G. Elliott [25]. The result of Pedersen supplied a powerful tool for the work of Elliott. The result of Elliott suggests an interesting problem: Find a separable infinite-dimensional simple C*-algebra which has only trivial central sequences.

In mathematical physics, one often meets unbounded derivations which are defined as infinitesimal generators of one-parameter groups of *-automorphisms on C*-algebras. Under some assumptions (for example, the semi-boundedness of the Hamiltonian) we may reduce the study of these unbounded derivations to one of bounded derivations. This was discovered by Borchers [4], (see also Dell'Antonio [19] and Arveson [3]). However, there are many important derivations in mathematical physics which do not satisfy the semi-boundedness (for example, the total energy of lattice systems). It is an important and challenging problem to study unbounded derivations in C*-algebras.

§2. Closability. Now suppose that $\mathcal{D}(\delta) \subsetneqq \mathfrak{A}$; then δ is not necessarily closable - in fact, Bratteli and Robinson [9] give an example of a C*-algebra which has a non-closable *-derviation. They proved [9] that if there is a sufficiently large family of states

(φ_α) such that $\varphi_\alpha(\delta(a)) = 0$ for $a \in \mathfrak{D}(\delta)$ and all α, then δ is closable. In particular, if $\mathfrak{A}$ is a simple C*-algebra (this assumption is often enough for C*-physics), δ is closable if there is a state φ such that for all $a \in \mathfrak{D}(\delta)$ $\varphi(\delta(a)) = 0$. Chi [16] proved that δ is closable if and only if $\mathfrak{D}(\delta^*)$ contains a sufficiently large family of linear functionals, where $\mathfrak{D}(\delta^*)$ is the domain of the adjoint δ^* of δ in $\mathfrak{A}^*$. In particular, if $\mathfrak{A}$ is a simple C*-algebra, then it is closable if and only if $\mathfrak{D}(\delta^*) \neq (0)$. These results suggest an interesting problem.

<u>Problem 1</u>. Suppose $\mathfrak{A}$ is a simple C*-algebra and δ is a closed *-derivation. Can we conclude that the norm closure of $\delta(\mathfrak{D}(\delta))$ is not $\mathfrak{A}$?

For commutative C*-algebras, the answer is negative - for example, if $\mathfrak{A} = C[0,1]$ and $\delta = \frac{d}{dx}$, then $\delta(\mathfrak{D}(\delta)) = C[0,1]$. The answer is obviously positive if δ is a bounded *-derivation. The answer is not known if δ is a bounded (not necessarily *-) derivation. However, Stampfli [69] proved that $\overline{\delta(B(\mathcal{H}))} \subsetneqq B(\mathcal{H})$ for every bounded derivation δ.

If $\mathfrak{A}$ has no identity and δ is closable, then adjoin an identity to $\mathfrak{A}$ and define $\delta(1) = 0$; then δ becomes a closable *-derivation in a C*-algebra with identity and moreover $1 \in \mathfrak{D}(\delta)$. On the other hand, Bratteli and Robinson [10] proved that if $\mathfrak{A}$ has an identity and δ is closed, then $1 \in \mathfrak{D}(\delta)$ (Chi [16] gave another nice proof to this fact). Therefore to study the closable derivations, it is enough to assume that $\mathfrak{A}$ has an identity and $\mathfrak{D}(\delta)$ contains it. Henceforth, we shall assume that $\mathfrak{A}$ has an identity and $\mathfrak{D}(\delta)$ contains it.

Powers and Sakai [61] proved that if the positive portion

of $\mathfrak{D}(\delta)$ is closed under the square root operation, then δ is closable.

The condition is obviously not necessary. However, it is often applicable to the C*-algebras appearing in quantum physics. The closability condition of Powers and Sakai is not strong enough to apply to general cases. With the help of a deep result of Cuntz [18] (a nice result of Chi [16] is also needed) Ôta [56] recently proved the following theorem.

Theorem 2.1. Let δ be a closed *-derivation in a C*-algebra. Suppose that the positive portion of $\mathfrak{D}(\delta)$ is closed under the square root operation; then $\mathfrak{D}(\delta) = \mathfrak{A}$ and so δ is bounded.

The closability condition of Powers and Sakai suggests another interesting problem. If $\mathfrak{D}(\delta) = \mathfrak{A}$, then the positive portion of $\mathfrak{D}(\delta)$ is closed under the square root operation, so that δ is closable. Since $\mathfrak{D}(\delta) = \mathfrak{A}$, δ is closed and hence, bounded by the closed graph theorem.

Now look at the general Banach algebra. Let $\mathfrak{L}$ be a semi-simple Banach algebra and let δ be a derivation with $\mathfrak{D}(\delta) = \mathfrak{L}$. Then Johnson and Sinclair [34] proved that δ is closed, so that it is bounded.

Problem 2. Can we formulate a closability condition of a densely defined derivation in a semi-simple Banach algebra which includes the theorem of Johnson and Sinclair as a special case?

Recently, Kishimoto [42] introduced the notion of dissipativity into the study of unbounded derivations. We shall define a slightly weaker condition than Kishimoto's.

<u>Definition 2.1</u>. A *-derivation δ in a C*-algebra $\mathfrak{A}$ is said to be well-behaved if for $x(>0) \in \mathfrak{D}(\delta)$, there is a state φ_x on $\mathfrak{A}$ such that $\varphi_x(x) = \|x\|$ and $\varphi_x(\delta(x)) = 0$.

<u>Theorem 2.2</u>. (Kishimoto [42]). If a *-derivation δ is well-behaved, then δ is closable and its closure is again well-behaved.

Proof. Suppose δ is not closable; then the closure F of the graph $\{(x,\delta(x)) | x \in \mathfrak{D}(\delta)\}$ in $\mathfrak{A} \oplus \mathfrak{A}$ contains an element $(0,a)\ (a \neq 0)$. Since δ is a derivation, F contains $\{(0,a) | a \in I\}$, where I is a non-zero two-sided closed ideal of $\mathfrak{A}$. Hence there is a sequence $\{x_n\}$ $(x_n^* = x_n)$ in $\mathfrak{D}(\delta)$ such that $x_n \to 0$ and $\delta(x_n) \to y\ (>0)$ with $\|y\| = 1$. Take a positive element u in $\mathfrak{D}(\delta)$ such that $\|u - y\| < 1/2$. For real λ, $\varphi_{u + \lambda x_n + |\lambda| \|x_n\|}(\delta(u + \lambda x_n + |\lambda| \|x_n\|)) = 0$. Take an accumulation point φ_λ of $\{\varphi_{u + \lambda x_n + |\lambda| \|x_n\|}\}$ in the state space; then $\varphi_\lambda(\delta(u) + \lambda y) = 0$ and $\varphi_\lambda(y) = \varphi_\lambda(u) - \varphi_\lambda(u-y) > 1 - 1/2 = 1/2$. Take $\lambda = 2\|\delta(u)\|$; then $\varphi_\lambda(\delta(u) + \lambda y) > 2\|\delta(u)\| \cdot 1/2 + \varphi_\lambda(\delta(u)) \geqq \|\delta(u)\| - \|\delta(u)\| = 0$, a contradiction. Hence δ is closable. Take $x \in \mathfrak{D}(\bar{\delta})$ $(x > 0)$ and let $x_n (x_n^* = x_n) \to x$ and $\delta(x_n) \to \bar{\delta}(x)$. Then $\|x_n\| + x_n \to \|x\| + x$ and $\delta(\|x_n\| + x_n) \to \bar{\delta}(x)$.
$\varphi_{\|x_n\| + x_n}(\|x_n\| + x_n) = \| \|x_n\| + x_n\|$ and $\varphi_{\|x_n\| + x_n}(\delta(x_n)) = 0$. Let φ_0 be an accumulation point of $\{\varphi_{\|x_n\| + x_n}\}$; then $\varphi_0(\|x\| + x) = \| \|x\| + x\| = 2\|x\|$ so that $\varphi_0(x) = \|x\|$. Moreover,

$\varphi_0(\overline{\delta}(x)) = 0$. Hence $\overline{\delta}$ is well-behaved. This completes the proof.

Corollary 2.1. ([42]). A *-derivation δ in $\mathfrak{A}$ is well-behaved if the positive portion of $\mathcal{D}(\delta)$ is closed under the square root operation.

Definition 2.2. A *-derivation δ in $\mathfrak{A}$ is said to be approximately inner if there is a sequence of self-adjoint elements (h_n) in $\mathfrak{A}$ such that

$$\delta(x) = \lim_n i\,[h_n, x] \quad (x \in \mathcal{D}(\delta)) .$$

Corollary 2.2. A *-derivation δ in $\mathfrak{A}$ is well-behaved if it is approximately inner.

For the proof, refer to the discussion on page 285 of [60].

Remark. If $\mathfrak{A} = C[0,1]$ and $\delta = \frac{d}{dx}$, then δ is not well-behaved. But if $\mathfrak{A}_0 =$ the algebra of all continuous functions f on $[0,1]$ with $f(0) = f(1)$ and $\mathcal{D}(\delta) =$ the algebra of all continuous differentiable functions g on $[0,1]$ with $g'(0) = g'(1)$ and $g(0) = g(1)$, then δ is well-behaved in $\mathfrak{A}_0$.

§3. Domains of closed derivations. In mathematical physics, unbounded derivations are often defined by Hamiltonians. In those cases, it is not difficult to see that the derivations are closable.

Let δ be a closed *-derivation in a C*-algebra $\mathfrak{A}$. Powers [59] initiated a study of the domain of δ . Introducing a clever operator calculus, he stated the following theorem:

(*) Let $a = a^* \in \mathcal{D}(\delta)$ and let f be a C^1-function (continuously differentiable) on the real line; then $f(a) \in \mathcal{D}(\delta)$ and $\|\delta(f(a))\| \leq \|f'\|_\infty \|\delta(a)\|$. However, Robinson [10] found an error in the proof, and Bratteli and Robinson [10] noted that the proof of Powers establishes the following theorem.

Theorem 3.1. If f is a function on the real line such that $|||\tilde{f}||| = \int^\infty |t\,\tilde{f}(t)|dt < +\infty$, where $\tilde{f}$ is the Fourier transform of f, then $f(a) \in \mathcal{D}(\delta)$ and $\|\delta(f(a))\| \leq |||\tilde{f}|||\ \|\delta(a)\|$.

In particular, if f is a C^2-function, then $f(a) \in \mathcal{D}(\delta)$.

In his dissertation, Chi [16] made an interesting observation. Let

$$\mathcal{L} = \{ \begin{pmatrix} a & \delta(a) \\ 0 & a \end{pmatrix} \mid a \in \mathcal{D}(\delta) \} ;$$

then $\mathcal{L}$ is a Banach algebra of bounded operators on a Hilbert space, since δ is closed and the mapping

$$a \to \begin{pmatrix} a & \delta(a) \\ 0 & a \end{pmatrix}$$

is an algebraic isomorphism of $\mathcal{D}(\delta)$ onto $\mathcal{L}$. Now let $a = a^* \in \mathcal{D}(\delta)$, and $T = \begin{pmatrix} a & \delta(a) \\ 0 & a \end{pmatrix}$; then we can show that the spectrum of T = the spectrum of a and

$$\|e^{itT}\| = O(|t|) \quad \text{as} \quad |t| \to \infty$$

([59]).

From the above considerations, if $f \in C^2(R)$, then we can define $f(T)$. If the first part of Power's statement (*) is true, then we can define $f(T)$ for $f \in C^1(R)$. There are some related developments in the spectral theory of non-self adjoint operators in a Hilbert space ([17],[40]). Let A be a bounded operator on a Hilbert space such that $\|e^{itA}\| = O(|t|^k)$ as $|t| \to \infty$ for some positive integer k ; then the spectum of A lies on the real line and for $f \in C^{k+2}(R)$, $f(A) = \int \tilde{f}(t)e^{itA}dt$, where $\tilde{f}$ is the Fourier transform of f and $\|f(A)\| \leqq M\int|\tilde{f}(t)|\,|t|^k dt$, where M is a constant. If $\|e^{itA}\| = O(1)$, then by Nagy's theorem [53] , we can define $f(A)$ for $f \in C^0(R)$. It has been conjectured that if $\|e^{itA}\| = O(|t|^k)$ as $|t| \to \infty$, $f(A)$ can be defined for $f \in C^{k+1}(R)$. This is correct in our case (for general k , see [16]). Moreover the question of whether or not $\|e^{itA}\| = O(|t|^k)$ as $|t| \to \infty$ implies the existence $f(A)$ for $f \in C^k(R)$ ([40]), has been asked. Therefore, if the first part of the statement (*) is not true, we have an example of a non-self adjoint operator on a Hilbert space which shows that spectral theory cannot be extended to C^1-functions.

In fact, McIntosh [52] constructed a closed *-derivation δ in $B(\mathcal{H})$ such that there is an element $a = a^* \in \mathcal{D}(\delta)$ and an $f \in C^1(R)$ such that $f(a) \notin \mathcal{D}(\delta)$. This is an interesting contribution of the theory of unbounded derivations to other parts of the subject.

The Power's proof of statement (*) is correct if $\mathfrak{A}$ is commutative - in fact, $\delta(f(a)) = f'(a)\delta(a)$ for $f \in C^1(R)$ and $a = a^* \in \mathcal{D}(\delta)$ (cf. [59]) .

It would be interesting to settle the following problem.

<u>Problem 3</u>. Let A be a bounded operator on a Hilbert space such that $\|e^{itA}\| = O(|t|)$ as $|t| \to \infty$ and $f(A)$ is definable for every $f \in C^1(R)$. Can we characterize such operators?

Problem 4. (Herman-Powers). Let $C[0,1]$ and let δ be a closed derivation in $C[0,1]$. Can we characterize δ ? (For example, $\delta = f \cdot \frac{d}{dx}$? , where f is some function on $[0,1]$).

This problem is interesting for the study of C*-differential manifolds.

§4. Generators. Let δ be a closable *-derivation in $\mathfrak{A}$. δ is said to be a pregenerator if its closure $\bar{\delta}$ is the (infinitesimal) generator of a strongly continuous one-parameter group of *-automorphisms on $\mathfrak{A}$.

An important problem is under what conditions δ is a pre-generator. It follows from semi-group theory that δ is a pregenerator if and only if $\|a + \lambda\delta(a)\| \geqq \|a\|$ $(a \in \mathfrak{D}(\delta))$ and the ranges $(\lambda \pm \delta)\mathfrak{D}(\delta)$ are dense in $\mathfrak{A}$ for all real $\lambda \neq 0$. However, for unbounded derivations, we can weaken these conditions.

Theorem 4.1. ([10], [60]). Let δ be a *-derivation in $\mathfrak{A}$. Then δ is a pre-generator if and only if δ is well-behaved and $(1 \pm \delta)\mathfrak{D}(\delta)$ are dense in $\mathfrak{A}$.

For the proof, see the discussions of pages 285-286 in [60]. The well-behavedness of δ implies $\|a + \lambda\delta(a)\| \geqq \|a\|$ $(a \in \mathfrak{D}(\delta))$.

Let δ_1 , δ_2 be two *-derivations in $\mathfrak{A}$. δ_1 is said to be an extension of δ if $\delta_1 \supset \delta$.

Theorem 4.2. Let δ be a well-behaved *-derivation in $\mathfrak{A}$. Then there is a maximal well-behaved *-derivation $\tilde{\delta}$ such that $\delta \subset \tilde{\delta}$.

The proof is very easy.

Let δ be a *-derivation in $\mathfrak{A}$. It is not so restrictive to assume that δ is well-behaved. However, the density of $(1 \pm \delta)\mathcal{D}(\delta)$ is not apparent in many interesting cases. The real problems often involve proving that density from available conditions.

Let $\mathcal{D} = \bigcap_{n=1}^{\infty} \mathcal{D}(\delta^n)$. An element $a \in \mathcal{D}$ is said to be analytic if there is a positive number such that

$$\sum_{n=0}^{\infty} \frac{\|\delta^n(a)\|}{n!} s^n < +\infty \quad .$$

Let $A(\delta)$ be the set of all analytic elements with respect to δ.

<u>Theorem 4.3</u> ([10],[60]). If δ is a well-behaved *-derivation in $\mathfrak{A}$ and $A(\delta)$ is dense, then $(1 \pm \delta)\mathcal{D}(\delta)$ is dense so that δ is a pregenerator.

Let δ be a generator. Then $\exp t\delta = \rho(t)$ ($\exp t\delta$ is the symbolic notation for the one-parameter group generated by an unbounded δ) is a strongly continuous one-parameter group of *-automorphisms on $\mathfrak{A}$. (Recall that δ is a *-derivation.)

Let $A_2(\delta)$ be the set of all entire analytic vectors a - i.e.,

$$\sum_{n=0}^{\infty} \frac{\|\delta^n(a)\|}{n!} r^n < +\infty$$

for all positive numbers r, and $A_2(\delta)$ the set of all geometric elements b - i.e., $\|\delta^n(b)\| \leq M_b^n \|b\|$, where M_b depends on b.

Then clearly $A(\delta) \supset A_1(\delta) \supset A_2(\delta)$, and each is a dense *-subalgebra of $\mathfrak{A}$ (cf. Bratteli and Robinson [10]).

Theorem 4.4 ([65]). Let p be a projection in $\mathfrak{A}$ and let ε be a positive number; then there exists a projection q in $A(\delta)$ such that $\|p - q\| < \varepsilon$.

Problem 5. For a projection $p \in \mathfrak{A}$ and a positive number $\varepsilon > 0$, can we find a projection $q \in A_1(\delta)$ (or $A_2(\delta)$) such that $\|p - q\| < \varepsilon$?

The proof of [65] does not apply to these cases. This problem seems to be important for the core problem (problem 14).

Let $a = a^* \in A(\delta)$; then we can define $\rho(z)(a) = \sum_{n=0}^{\infty} \frac{\delta^n(a)}{n!} z^n$ ($z \in C$ with $|z| < s(a)$) , where $s(a)$ is a positive number depending on a . $\rho(z)(ab) = \rho(z)(a)\rho(z)(b)$ for $a,b \in A(\delta)$ if $|z| < \min(s(a),s(b))$. In particular, if $a \in A_1(\delta)$, then $\rho(z)(a)$ can be defined for all $z \in C$, and $\rho(z)$ is a strongly continuous complex parameter group of automorphisms on $A_1(\delta)$.

If $p \in A(\delta)$ is a projection, then $\rho(z)(p)$ is an idempotent, so that Spectrum $(\rho(z)(p))$ = Spectrum (p) . Generally the spectrum of $\rho(z)(a)$ will depend on z . In fact, let $C_0(R)$ be the C*-algebra of all continuous functions on the real line vanishing at infinity. Let $\delta = \frac{d}{dt}$; then $(\exp t\,\delta)f(x) = f(x+t)$ ($f \in C_0(R)$) , where $x,t \in R$. Let $f(x) = \frac{1}{1+x^2}$; then $f \in A(\delta)$

and Spectrum (f) consists of positive numbers. On the other hand,

$$(\exp z\ \delta)f(x) = \sum_{n=0}^{\infty} \frac{z^n f^{(n)}(x)}{n!} = \frac{1}{1+(x+z)^2}$$

is complex-valued. However, if $A(\delta)$ contains many projections, the non-invariance of the spectrum under $\rho(z)$ is not clear. Later we shall discuss this problem in the case of uniformly hyperfinite C*-algebras.

In mathematical physics, we are often concerned with a C*-algebra $\mathfrak{A}$ containing an identity 1 and an increasing sequence of C*-subalgebras $\{\mathfrak{A}_n\}$ of $\mathfrak{A}$ such that $1 \in \mathfrak{A}_n$ and the uniform closure of $\bigcup_{n=1}^{\infty} \mathfrak{A}_n$ is $\mathfrak{A}$. In addition, we are given a *-derivation δ in $\mathfrak{A}$ satisfying the following conditions: (1) $\mathcal{D}(\delta) = \bigcup_{n=1}^{\infty} \mathfrak{A}_n$; (2) there is a sequence of self-adjoint elements (h_n) in $\mathfrak{A}$ such that $\delta(a) = i[h_n, a]\ (a \in \mathfrak{A}_n)$.

<u>Definition 4.1</u>. We shall call such a derivation δ a normal *-derivation in $\mathfrak{A}$.

<u>Definition 4.2</u>. Let $\{\rho(t)\}$ be a strongly continuous one-parameter group of *-automorphisms on $\mathfrak{A}$. $\{\rho(t)\}$ is said to be approximately inner if there is a sequence of uniformly continuous one-parameter groups $\{\rho_n(t)\}$ of inner *-automorphisms on $\mathfrak{A}$ such

that $\|\rho_n(t)(a) - \rho(t)(a)\| \to 0$ uniformly on every compact subset of R and for each fixed $a \in \mathfrak{A}$. (By using the category theorem, we can easily see that $\|\rho_n(t)(a) - \rho(t)(a)\| \to 0$ (simple convergence) implies the uniform convergence of $\|\rho_n(t)(a) - \rho(t)(a)\| \to 0$ on every compact subset of R.) In this case, we say that the C*-dynamics $\{\mathfrak{A}, \rho(t)\}$ is an approximately inner dynamics.

<u>Theorem 4.5</u> ([37]). Suppose that δ is a normal *-derivation in a C*-algebra $\mathfrak{A}$ such that $\delta(a) = i[h_n, a]$ $(a \in \mathfrak{A}_n)$. If $h_n \in \mathfrak{A}_{n+1}$ and there is an element k_n in $\mathfrak{A}$ such that $\|h_n - k_n\| = O(n)$, then δ is a pre-generator, the C*-dynamics $\{\mathfrak{A}, \exp t\bar{\delta}\}$ is approximately inner, and $\exp t\bar{\delta}(a) = \lim_n \exp t\,\delta_{ih_n}(a)$ $(a \in \mathfrak{A})$.

The following theorem is essentially due to Kishimoto [42]; but Jørgensen [35] extended it to the form in which it is stated.

<u>Theorem 4.6</u>. Suppose that δ is a normal *-derivation in $\mathfrak{A}$ such that $\delta(a) = i[h_n, a]$ $(a \in \mathfrak{A}_n)$ and $\mathfrak{A}_n$ is finite-dimensional for all n. Let P_n be a conditional expectation of $\mathfrak{A}$ onto $\mathfrak{A}_n$. (Recall that $\mathfrak{A}_n$ is a finite-dimensional C*-algebra.) If $\|h_n - P_n(h_n)\| = O(1)$, then δ is a pregenerator, the dynamics is approximately inner, and $\exp t\,\delta(a) = \lim_n \exp \delta_{ih_n}(a)$ $(a \in \mathfrak{A})$.

It would be interesting to settle the following problem.

<u>Problem 6</u>. Let L be a densely defined dissipative linear operator in a Banach space E. Let $\{V_n\}$ be an increasing sequence of closed subspaces in E such that $\bigcup_{n=1}^{\infty} V_n \subset \mathcal{D}(L)$

and $\bigcup_{n=1}^{\infty} V_n$ is dense in E. Suppose that there is a bounded linear operator L_n of V_n into V_n such that

$$\sup_{\substack{\|x\| \leq 1 \\ x \in V_n}} \|L(x) - L_n(x)\| = O(1) .$$

Can we conclude that L is a pregenerator of a contraction semi-group? (Note that since L is closed, $L|V_n$ is a bounded linear operator of V_n into E.)

Definition 4.3. A normal *-derivation δ in $\mathfrak{A}$ is said to be commutative if we can choose the sequence (h_n) such that $h_n h_m = h_m h_n$ $(m,n=1,2,\ldots)$ (see Definition 4.1).

Theorem 4.7 ([66]). If δ is commutative, then it has an extension δ_1, such that δ_1 is a generator, the C*-dynamics $\{\mathfrak{A}, \exp t\,\delta_1\}$ is approximately inner, and $\exp t\,\delta_1(a) = \lim_n \exp t\,\delta_{ih_n}(a)$ $(a \in \mathfrak{A})$.

§5. Ground States. We introduce a class of states of some importance in quantum physics.

Definition 5.1. Let $\{\mathfrak{A}, \rho(t)\}$ be a C*-dynamics, and let δ be the generator of $\{\rho(t)\}$. A state φ on $\mathfrak{A}$ is said to be a ground state for $\{\rho(t)\}$ if $-i\varphi(a^*\delta(a)) \geqq 0$ for $a \in \mathcal{D}(\delta)$.

A ground state φ is invariant under $\{\rho(t)\}$ - i.e., $\varphi(\rho(t)(a)) = \varphi(a)$ $(a \in \mathfrak{A})$ ([60]).

<u>Theorem 5.1</u>. ([60]). Let $\{\mathfrak{A}, \rho(t)\}$ be an approximately inner C*-dynamics; then it has a ground state.

Lance and Niknam [46] constructed an example of a C*-dynamics without a ground state.

If φ is a ground state of $\{\mathfrak{A}, \rho(t)\}$, $\{\pi_\varphi, \mathfrak{H}_\varphi\}$ is the GNS *-representation of $\mathfrak{A}$ constructed from φ, and $U_\varphi(t)a_\varphi = (\rho(t)(a))_\varphi$ $(a \in \mathfrak{A})$; then $U_\varphi(t)$ can be uniquely extended to a unitary operator (denoted, again, by $U_\varphi(t)$) on $\mathfrak{H}_\varphi$. $t \to U_\varphi(t)$ is a strongly-continuous, one-parameter group of unitary operators and $U_\varphi(t)\pi_\varphi(a)U_\varphi(-t) = \pi_\varphi(\rho(t)(a))$ $(a \in \mathfrak{A})$.

By Stone's theorem $U_\varphi(t) = \exp it H_\varphi$ with H_φ self adjoint. Since $-i\varphi(a^*\delta(a)) \geqq 0$ $(a \in \mathcal{D}(\delta))$; $H_\varphi \geqq 0$. By Borchers' theorem [4], the semi-boundedness of H_φ implies that there is a strongly-continuous, one-parameter group $\{V(t)\}$ with $V(t) \in \overline{\pi_\varphi(\mathfrak{A})}$ such that $V(t)\pi_\varphi(a)V(-t) = U_\varphi(t)\pi_\varphi(a)U_\varphi(-t)$, $a \in \mathfrak{A}$. Let M be the W*-algebra generated by $\{\pi_\varphi(\mathfrak{A}), U_\varphi(R)\}$ and let B' be an element in M'; then

$$\varphi_{B'}(a) = \frac{\langle \pi_\varphi(a)B'1_\varphi, B'1_\varphi\rangle}{\langle B'1_\varphi, B'1_\varphi\rangle}$$

$(a \in \mathfrak{A})$ is also a ground state for $\{\rho(t)\}$.

Now let $\mathcal{S}_G$ be the set of all ground states on $\mathfrak{A}$ for $\{\rho(t)\}$. Then $\mathcal{S}_G$ is a compact convex subset of the state space $\mathcal{S}$ of $\mathfrak{A}$. If φ is an extreme point in $\mathcal{S}_G$, then, from the above considerations, $M = B(\mathcal{H}_\varphi)$, where $B(\mathcal{H}_\varphi)$ is the algebra of all bounded operators on $\mathcal{H}_\varphi$. Since $V(t)U_\varphi(-t) \in M' = (\lambda 1)$, $M = \overline{\pi_\varphi(\mathfrak{A})}$ and so φ is a pure state on $\mathfrak{A}$. Hence we have:

<u>Theorem 5.2</u>. If $\mathcal{S}_G$ is the set of all ground states for $\{\mathfrak{A}, \rho(t)\}$ then all extreme points in $\mathcal{S}_G$ are pure states.

<u>Definition 5.2</u>. A ground state φ for a C*-dynamics $\{\mathfrak{A}, \rho(t)\}$ is said to be a physical ground state if the representation $\{\pi_\varphi, U_\varphi, \mathcal{H}_\varphi\}$ constructed via φ satisfies the following condition: $K_\varphi = \{\xi | H_\varphi \xi = 0, \xi \in \mathcal{H}_\varphi\}$ is one-dimensional, where $U_\varphi(t) = \exp it H_\varphi$.

The vacuum is a physical ground state. It is a state of 0 energy and momentum in quantum field theory. From the above considerations, a physical ground state must be a pure state. But a pure, ground state is not necessarily a physical ground state (even in the finite-dimensional case).

<u>Problem 7</u>. Can we characterize those approximately inner C*-dynamics that possess a physical ground state?

Problem 8. Can we characterize those C*-dynamics that possess a physical ground state?

Problem 9. Can we characterize those C*-dynamics $\{\mathfrak{A}, \rho(t)\}$ in which all pure ground states are physical ground states?

If a C*-dynamics $\{\mathfrak{A}, \rho(t)\}$ is $\rho(R)$ - abelian (in particular, $\rho(R)$ - asymptotically abelian), then all pure ground states are physical ground states (cf. [22], [23], [47]).

If a C*-dynamics has a unique ground state, then it must be a physical ground state.

If a C*-dynamics has two different physical ground states, then they are centrally orthogonal.

Problem 10. Is there a C*-dynamics $\{\mathfrak{A}, \rho(t)\}$ with a simple C*-algebra with identity having at least two different physical ground states?

Gross [27] gave a method for proving existence of physical ground states which is applicable to a wide variety of quantum field theoretic models.

For a ground state φ for $\{\mathfrak{A}, \rho(t)\}$, consider the representation $\{\pi_\varphi, U_\varphi, \mathcal{H}_\varphi\}$. Let $V(t) = \exp it\, h_\varphi$, where $V(t) \in \overline{\pi_\varphi(\mathfrak{A})}$; then $h_\varphi\ \eta\ \overline{\pi_\varphi(\mathfrak{A})}$, so that there is a sequence of self adjoint elements (h_n) in $\mathfrak{A}$ such that

$$\exp it\, \pi_\varphi(h_n)\pi_\varphi(a)\exp - it\, \pi_\varphi(h_n) \to \pi_\varphi(\rho(t)(a))$$

in the strong operator topology in $\mathcal{H}_\varphi$. This implies that $\{\rho(t)\}$ is approximately inner in a weak sense.

The following problem poses itself.

Problem 11. Let $\{\mathfrak{A}, \rho(t)\}$ be a C*-dynamics with a separable simple C*-algebra $\mathfrak{A}$ with identity, and suppose that it has a ground state. Can we conclude that $\{\mathfrak{A}, \rho(t)\}$ is an approximately inner dynamics?

Problem 11'. Characterize those C*-dynamics possessing a ground state.

These problems may not be totally intractable - in fact, Olesen and Pedersen [55] gave two conditions which are necessary and sufficient for H to be positive in the universal representation of $\mathfrak{A}$ (this is too strong for the physical setting). One of them is that the dynamics to be approximately inner.

As the first step, one may assume that a ground state is unique for the dynamics.

Problem 11". Characterize those C*-dynamics which possess a unique ground state.

This problem is quite important. Even under the added assumption that the C*-algebra $\mathfrak{A}$ is simple, norm-separable, and has an identity - also assuming G-abelianness or asymptotic abelianness (cf. [22],[23],[47]) with respect to translation or time evolution, although lattice systems(Ising models, Heisenberg models) do not satisfy the asymptotic abelianness with respect to time evolution. The case in which the dynamics is approximately inner, is of paramount interest.

§6. KMS states. We introduce another class of states on a C*-algebra which is important in quantum physics.

<u>Definition 6.1</u>. Let $\{\mathfrak{A}, \rho(t)\}$ be a C*-dynamics. For a a state φ_β on $\mathfrak{A}$ is said to be a KMS state for $\{\rho(t)\}$ at inverse temperature β if for $a,b \in \mathfrak{A}$, there is a bounded continuous function $F_{a,b}$ on the strip $0 \leq \mathrm{Im}(z) \leq \beta$ (or $0 \geq \mathrm{Im}(z) \geq \beta$) in the complex plane which is analytic on $0 < \mathrm{Im}(z) < \beta$ (or $0 > \mathrm{Im}(z) > \beta$) so that $F_{a,b}(t) = \varphi_\beta(a\rho(t)(b))$ and $F_{a,b}(t+i\beta) = \varphi_\beta(\rho(t)(b)a)$.

The KMS condition gives every evidence of being the abstract formulation of the condition for equilibrium of a state [28]. A KMS state for $\{\rho(t)\}$ is invariant under $\{\rho(t)\}$. The set $\mathfrak{S}_{K,\beta}$ of all KMS states at β on $\mathfrak{A}$ is a compact convex subset of the state space $\mathfrak{S}$; and extreme points of $\mathfrak{S}_{K,\beta}$ are factorial states.

<u>Theorem 6.1</u> ([60]). Suppose that $\{\mathfrak{A}, \rho(t)\}$ is approximately inner and $\mathfrak{A}$ has a tracial state; then it has a KMS state φ_β for $\{\rho(t)\}$ at every inverse temperature β $(-\infty < \beta < +\infty)$.

<u>Theorem 6.2</u> ([45]). Suppose that $\{\mathfrak{A}, \rho(t)\}$ is approximately inner and there is a sequence of one parameter groups $\{\exp it\, \delta_{ih_n}\}$ $(h_n \in \mathfrak{A})$ of inner *-automorphisms on $\mathfrak{A}$ such that $\|\rho(t)(a) - \exp t\, \delta_{ih_n}(a)\| \to 0 (n\to\infty)$ for each $a \in \mathfrak{A}$ and $\rho(t)(h_n) = h_n (n=1,2,\ldots)$. Then if $\{\rho(t)\}$ has a KMS state φ_{β_0} at some β_0 ; it has a KMS state φ_β at every β $(-\infty < \beta < +\infty)$.

Remark. If the generator of $\{\rho(t)\}$ has a commutative derivation as a core, then it satisfies all the conditions in Theorem 6.2 (cf. Theorem 4.7).

Theorem 6.3 ([36]). Suppose that $\{\mathfrak{A}, \rho(t)\}$ is an approximately inner dynamics and $\|\exp t\, \delta_{ih_n}(a) - \rho(t)(a)\| \to 0$ $(n \to \infty)$ for each $a \in \mathfrak{A}$ $(h_n = h_n^* \in \mathfrak{A})$ and $\|\exp t\, \delta i(h_m - h_n)(a)\| \to 0$ $(m,n \to \infty)$ $(a \in \mathfrak{A})$. Then if $\{\rho(t)\}$ has a KMS state at some β_0 ; it has a KMS state at every β $(-\infty < \beta < +\infty)$.

Remark. If the generator of $\{\rho(t)\}$ is the closure of a normal *-derivation in $\mathfrak{A}$, then all the conditions in Theorem 6.3 are satisfied.

Problem 12. Can we omit the condition $\|\exp t\, \delta i(h_m - h_n)\| \to 0$ $(m,n \to \infty)$ $(a \in \mathfrak{A})$ in Theorem 6.3?

Lance and Niknam [46] have constructed an example of a C*-dynamics $\{\mathfrak{A}, \rho(t)\}$ with a separable simple C*-algebra with identity and a tracial state τ which does not have a ground state. Since a tracial state is a KMS state at $\beta = 0$ for every $\{\rho(t)\}$; the existence of a KMS state does not imply that the dynamics is approximately inner. However, the existence of KMS states $(\varphi_{\beta n})$ $(\beta n \uparrow \infty)$, at least, implies the existence of a ground state (cf. the next section), which suggests the following problem.

Problem 13. Is there a C*-dynamics $\{\mathfrak{A}, \rho(t)\}$ with a separable simple C*-algebra $\mathfrak{A}$ with identity such that $\{\rho(t)\}$ has a KMS state at some positive β, but which is not approximately inner (or has no ground state)?

§7. UHF-algebras. In this section, we shall assume that $\mathfrak{A}$ is a uniformly hyperfinite C*-algebra (UHF algebra). Such algebras are important in quantum lattice systems and Fermion field theory.

Theorem 7.1 ([65]). Let $\{\mathfrak{A}, \rho(t)\}$ be a C*-dynamics with a UHF algebra $\mathfrak{A}$. Then there is an increasing sequence $\{\mathfrak{A}_n\}$ of finite type I subfactors in $\mathfrak{A}$ such that $1 \in \mathfrak{A}_n$, $\bigcup_{n=1}^{\infty} \mathfrak{A}_n$ is dense in $\mathfrak{A}$ and every element of $\bigcup_{n=1}^{\infty} \mathfrak{A}_n$ is analytic with respect to $\{\rho(t)\}$.

If δ is the infinitesimal generator of $\{\rho(t)\}$, there is a sequence of self-adjoint elements (h_n) in $\mathfrak{A}$ such that h_n is analytic for $\{\rho(t)\}$ and $\delta(a) = i[h_n, a]$ $(a \in \mathfrak{A}_n,\ n=1,2,\dots)$ (cf. [24],[60],[65]). Therefore the restriction δ_0 of δ to $\bigcup_{n=1}^{\infty} \mathfrak{A}_n$ is normal, so that it is well-behaved.

If $(1 - \delta_0)\mathfrak{A}_0$ $(\mathfrak{A}_0 = \bigcup_{n=1}^{\infty} \mathfrak{A}_n)$ is dense in $\mathfrak{A}$, then $\|\rho(t)(a) - \exp t\,\delta_{ih_n}(a)\| \to 0$ uniformly on every compact set of $(0, +\infty)$ for each $a \in \mathfrak{A}$.

On the other hand,

$$\|\rho(-t)(a) - \exp -t\,\delta_{ih_n}(a)\|$$

$$= \|\exp - t\,\delta_{ih_n}(\exp t\,\delta_{ih_n} - \rho(t))\,\rho(-t)(a)\|$$

$$\to 0 \quad (n \to \infty) \quad \text{for every} \quad t \in (0,\infty)\ .$$

Hence $\|\rho(t)(a) - \exp t\,\delta_{ih_n}(a)\| \to 0 \quad (n \to \infty)$ for each $t \in (-\infty,\infty)$.

The following problem is crucial.

Problem 14 (The Core Problem). Let $\{\mathfrak{A}, \rho(t)\}$ be a C*-dynamics with a UHF algebra $\mathfrak{A}$. Can we choose an increasing sequence $\{\mathfrak{A}_n\}$ of finite type I subfactors in $\mathfrak{A}$ such that $\mathfrak{A}_n \subset \mathfrak{D}(\delta)$ and $(1 - \delta)\,(\bigcup_{n=1}^{\infty} \mathfrak{A}_n)$ is dense in $\mathfrak{A}$, where δ is the generator of $\{\rho(t)\}$?

The following conjecture is plausible.

Conjecture. Any C*-dynamics with a UHF algebra $\mathfrak{A}$ is approximately inner.

The following theorem may be considered a positive step toward the proof of this conjecture.

Theorem 7.2 ([49]). Let $\{\mathfrak{A}, \rho(t)\}$ be a C*-dynamics with a UHF algebra $\mathfrak{A}$; then there exists a sequence $u_t^{(n)} \in \mathfrak{A}$ of norm-differentiable, unitary-valued $\rho(t)$-cocycles $u_{t+s}^{(n)} = u_t^{(n)}\rho(t)(u_s^{(n)})$ for $s, t \in R$ such that for fixed $a \in \mathfrak{A}$ $\|u_t^{(n)*} a\, u_t^{(n)} - \rho(t)(a)\| \to 0$ uniformly on R .

Let $A(\delta)$ be the set of all analytic elements with respect to $\rho(t) = \exp t\,\delta$ in a UHF algebra $\mathfrak{A}$. For each $a \in A(\delta)$, there is a positive number $s(a)$ such that

$$\sum_{n=0}^{\infty} \frac{\|\delta^n(a)\|}{n!} |z|^n < +\infty \quad (|z| < s(a)) .$$

Define $\rho(z)(a) = \sum_{n=0}^{\infty} \frac{\delta^n(a)}{n!} z^n$ $(|z| < \rho(a),\ z \in C)$. Then $\rho(z)(ab) = \rho(z)(a)\rho(z)(b)$ $(|z| < \min(s(a), s(b)))$. For any projection $p \in A(\delta)$, Spectrum $(\rho(z)(p))$ = Spectrum (p), so that Spectrum $(\rho(z)(a))$ = Spectrum (a) for a $(a^* = a) \in \bigcup_{n=1}^{\infty} \mathfrak{A}_n$. $(|z| < s(a))$.

<u>Problem 15</u>. Can we conclude Spectrum $(\rho(z)(a))$ = Spectrum (a) for $a(= a^*) \in A(\delta)$ $(|z| < s(a))$?

<u>Remark</u>. The problem has a negative answer in the general case, as mentioned before.

<u>Problem 16</u>. If problem 15 has a negative answer under what conditions can we conclude that $\{\rho(z)\}$ preserve the spectrum of $A(\delta)$?

<u>Problem 17</u>. Can we choose an increasing sequence $\{\mathfrak{A}_n\}$ of finite type I subfactors in $A(\delta)$ such that there exists a positive number s_0 for which $\bigcup_{n=1}^{\infty} \mathfrak{A}_n$ is dense in $\mathfrak{A}$ and

$$\sum_{n=1}^{\infty} \frac{\|\delta^n(a)\|}{n!} s_0 < +\infty \quad \text{for } a \in \bigcup_{n=1}^{\infty} \mathfrak{A}_n .$$

The proof of [65] does not guarantee the existence of an increasing sequence such as $\{\mathfrak{A}_n\}$. On the other hand, in quantum lattice models, we can often choose such a fixed positive number ([63]).

Now we shall define a normal *-derivation in UHF algebras more restrictively than in general cases.

Definition 7.1. Let δ be a *-derivation in a UHF algebra $\mathfrak{A}$. δ is said to be normal if there is an increasing sequence $\{\mathfrak{A}_n\}$ of finite type I subfactors in $\mathfrak{A}$ such that $\mathcal{D}(\delta) = \bigcup_{n=1}^{\infty} \mathfrak{A}_n$.

From Theorem 7.1, one sees that normal *-derivations in a UHF algebra play a key role. Most unbounded derivations in quantum lattice systems and Fermion field theory have normal *-derivations as their cores.

Let $\mathfrak{A}(\mathcal{H})$ be the canonical anti-commutation relation algebra over a separable Hilbert space $\mathcal{H}$ ([58]). Let S be a symmetric operator in $\mathcal{H}$ and put $\delta_S a(f) = i\, a(Sf)$ $(f \in \mathcal{D}(S))$, where $f \to a(f)$ is the basic linear isometric mapping of $\mathcal{H}$ into $\mathfrak{A}(\mathcal{H})$ which defines $\mathfrak{A}(\mathcal{H})$ canonically. Then δ_S extends (uniquely) to a *-derivation on a *-subalgebra $\mathfrak{A}_0(\mathcal{H})$ of $\mathfrak{A}(\mathcal{H})$ generated by $\{a(f) \mid f \in \mathcal{D}(S)\}$. So $K(\subset \mathcal{D}(S))$ is a finite-dimensional subspace of $\mathcal{H}$; then $\mathfrak{A}(K)$ is a finite type I subfactor of $\mathfrak{A}(\mathcal{H})$ and there is a self-adjoint element h in $\mathfrak{A}(\mathcal{H})$ such that $\delta_S(x) = i[h,x]$ $(x \in \mathfrak{A}(K))$. From this, we can easily conclude that δ_S is well-behaved. Hence δ_S is closable. The closure of δ_S is called the quasi-free *-derivation induced by

S and is denoted by δ_S again. Clearly $\delta_S = \delta_{\bar{S}}$.

Suppose that S is a closed symmetric operator and $S = UH$ is its polar decomposition. Since $\mathcal{H}$ has an increasing sequence $\{V_n\}$ of finite-dimensional subspaces such that $\bigcup_{n=1}^{\infty} V_n$ is a core for H ([51]) , $\bigcup_{n=1}^{\infty} V_n$ is also a core for S . From this we deduce easily.

<u>Theorem 7.3.</u> δ_S is the closure of a normal *-derivation in $\mathfrak{A}(\mathcal{H})$.

<u>Problem 18.</u> Does every closed *-derivation in a UHF algebra have a normal *-derivation as a core? (This problem was raised by Powers.)

<u>Remark.</u> Every closed *-derivation in a UHF algebra $\mathfrak{A}$ has an increasing sequence $\{\mathfrak{A}_n\}$ of finite type I subfactors in its domain such that $\bigcup_{n=1}^{\infty} \mathfrak{A}_n$ is dense in $\mathfrak{A}$ (cf. [9],[65]) .

<u>Theorem 7.4</u> ([51]). δ_S is a bounded derivation if and only if S is a bounded operator of trace class and in this case $\|\delta_S\| = \|S\|_1$, where $\|S\|_1$ is the trace norm.

Now let δ be a normal *-derivation in a UHF algebra $\mathfrak{A}$ and let (h_n) be a sequence of self-adjoint elements in $\mathfrak{A}$ such that $\delta(a) = i[h_n, a]$ $(a \in \mathfrak{A}_n)$ $(n = 1,2,\ldots)$, where $\mathcal{D}(\delta) = \bigcup_{n=1}^{\infty} \mathfrak{A}_n$. Suppose that $(1 \pm \delta)\mathcal{D}(\delta)$ are dense in $\mathfrak{A}$ then δ is a pre-generator and $\|\exp t\, \delta_{ih_n}(a) - \exp t\, \bar{\delta}(a)\| \to 0$ $(a \in \mathfrak{A})$ ([60]) .

For real β , let $\varphi_{\beta,n}(x) = \dfrac{\tau(xe^{-\beta h_n})}{\tau(e^{-\beta h_n})}$ $(x \in \mathfrak{A})$, where τ is the unique tracial state on $\mathfrak{A}$. If φ_β is an accumulation point of $\{\varphi_{\beta,n}\}$ in the state space $\mathfrak{S}$ of $\mathfrak{A}$, then φ_β is a KMS state for $\{\rho(t)\}$ at inverse temperature β ([60]).

If $h_n \geq 0$ and not invertible and φ is a state on $\mathfrak{A}$ such that $\varphi_n(h_n) = 0$, then each accumulation point of $\{\varphi_n\}$ in $\mathfrak{S}$ is a ground state for $\{\rho(t)\}$ ([60]).

<u>Theorem 7.5</u>. Let $\beta_n \uparrow \infty$ and let $\{\varphi_{\beta n}\}$ be a sequence of KMS states for $\{\rho(t)\}$ at inverse temperatures β_n, in a C*-dynamics $\{\mathfrak{A}, \rho(t)\}$. Then any accumulation point φ of $\{\varphi_{\beta n}\}$ is a ground state for $\{\rho(t)\}$.

<u>Proof</u>.

$$F_{a,b,n}(t) = \varphi_{\beta n}(a\rho(t)(b))$$

$$F_{a,b,n}(t+i\beta_n) = \varphi_{\beta n}(\rho(t)(b)a) \qquad (a,b \in \mathfrak{A}) .$$

and

$$|F_{a,b,n}(z)| \leqq \|a\| \, \|b\| \quad \text{on} \quad 0 \leqq I_m(z) \leqq \beta_n .$$

For $a,b \in \mathfrak{D}(\delta)$,

$$F'_{a,b,n}(t) = \varphi_{\beta n}(a\rho(t)(\delta(b)))$$

Hence

$$|F'_{a,b,n}(t)| \leqq \|a\| \, \|\delta(b)\| .$$

Therefore, from the theory of analytic functions, $\{F_{a,b,n}\}$ has a subsequence $\{F_{a,b,n_j}\}$ such that $\{F_{a,b,n_j}\}$ converges uniformly on compact subsets to a bounded holomorphic function $F_{a,b}$ on the upper half-plane and $F_{a,b}$ is continuous on $I_m(z) \geqq 0$. Moreover, at β, then we say that $\{\rho(t)\}$ has phase transition at β.

Let δ be a normal *-derivation in a UHF algebra $\mathfrak{A}$ with $\mathcal{D}(\delta) = \bigcup_{n=1}^{\infty} \mathfrak{A}_n$. Let P_n be the canonical conditional expectation of $\mathfrak{A}$ onto $\mathfrak{A}_n$ such that $\tau(xa) = \tau(P_n(x)a)$ $(a \in \mathfrak{A}_n)$, where τ is the unique tracial state on $\mathfrak{A}$.

Let (h_n) be a sequence of self-adjoint elements in $\mathfrak{A}$ such that $\delta(a) = i[h_n, a]$ $(a \in \mathfrak{A}_n)$ $(n = 1, 2, \ldots)$.

Definition 8.2. A normal *-derivation δ in a UHF algebra $\mathfrak{A}$ is said to satisfy the approximate boundedness condition if we can choose (h_n) such that $\|h_n - P_n(h_n)\| = O(1)$.

Suppose that δ satisfies the approximate boundedness condition. Then, by Kishimoto's theorem (Th. 4.6), δ is a pre-generator. Put $\rho(t) = \exp t\bar{\delta}$. Then $\{\rho(t)\}$ is approximately inner; and, for fixed $a \in \mathfrak{A}$, $\|\rho(t)(a) - \exp t\,\delta_{ih_n}(a)\| \to 0 \ (n \to \infty)$.

Theorem 8.1 ([1]). If a normal *-derivation δ in a UHF algebra $\mathfrak{A}$ satisfies the approximate boundedness condition, then the C*-dynamics $\{\mathfrak{A}, \exp t\bar{\delta}\}$ has no phase transition at every β $(-\infty < \beta < +\infty)$.

Remark. This theorem was first proved in [67],[68] for commutative derivations.

$F_{a,b}(t) = \varphi(a\rho(t)(b))$.

Since a KMS state is invariant under $\{\rho(t)\}$, φ is invariant under $\{\rho(t)\}$. Let $\{\pi_\varphi, U_\varphi, \mathcal{H}_\varphi\}$ be the *-representation of $\mathfrak{A}$ and $\{\rho(t)\}$ via φ , then $U_\varphi(t) = \exp itH$. $F_{a,b}(z) = \langle a^*_\varphi , \exp izHb_\varphi \rangle = \langle a^*_\varphi , \exp ixH \exp -yHb_\varphi \rangle$, where $z = x + iy$. It follows easily that $H \geq 0$. This completes the proof.

Problem 19. Suppose that a pure ground state φ is an accumulation point of $\{\varphi_{\beta n}\}$ in Theorem 7.5 ; then can we conclude that φ is a physical ground state?

A ground state may be considered a KMS state at infinite inverse temperature. On the other hand, a tracial state is a KMS state at zero inverse temperature.

Problem 20. Let $\{\mathfrak{A}, \rho(t)\}$ be a C*-dynamics with a simple C*-algebra $\mathfrak{A}$ with identity. Suppose that $\{\rho(t)\}$ has a tracial state and a ground state. Can we conclude that $\{\rho(t)\}$ has a KMS state at arbitrary β (or $\beta > 0$) ?

§8. Phase transition.

Definition 8.1. Let $\{\mathfrak{A}, \rho(t)\}$ be a C*-dynamics. Suppose that $\{\rho(t)\}$ has a KMS state φ_β at every β $(-\infty < \beta < +\infty)$. If $\{\rho(t)\}$ has only one KMS state φ_β at β , then we say that $\{\rho(t)\}$ has no phase transition at β . If $\{\rho(t)\}$ has at least two KMS states

Theorem 8.2 ([51]). Let H be a bounded self-adjoint operator with finite multiplicity on a Hilbert space $\mathfrak{H}$ - i.e., the W*-algebra generated by H in $\mathfrak{H}$ is a finite direct sum of finite copies of a maximal commutative *-algebra. Then the quasi-free *-derivation δ_H in $\mathfrak{A}(\mathfrak{H})$ induced by H is the closure of a normal *-derivation which satisfies the approximate boundedness condition. Consequently the quasi-free C*-dynamics $\{\mathfrak{A}(\mathfrak{H}), \exp t\,\delta_H\}$ has no phase transition at β $(-\infty < \beta < +\infty)$.

Remark. It is known that every quasi-free dynamics has no phase transition at β $(-\infty < \beta < +\infty)$. However, the proof of this Fermion field case is completely different from the proof of the quantum lattice system with bounded surface energy. The approximate boundedness condition now supplies a unified proof for both cases.

By Weyl's theorem, any self-adjoint H in $\mathfrak{H}$ can be written as $H = H_1 + K$, where H_1 is a diagonalizable self-adjoint operator and K is of Hilbert-Schmidt class. If K is of trace class, then by Theorem 7.4, we easily see that δ_H is the closure of a normal *-derivation satisfying the approximate boundedness condition. However, there is a self-adjoint operator which cannot be written as $H = H_1 + K$ with K of trace class. This gives rise to the following problem (and points to the need for further study of the decomposition problem for self-adjoint operators).

Problem 21. Suppose H is self-adjoint in a separable Hilbert space $\mathfrak{H}$. Can we conclude that δ_H is the closure of a normal *-derivation of $\mathfrak{M}(\mathfrak{H})$ satisfying the approximate boundedness condition?

Phase transition theory is one of most important branches in statistical mechanics. It has numerous scientific applications. However, the theory is far from complete and presents many theoretical puzzles.

We have seen that "no phase transition" is assured by the fairly general condition of "approximate boundedness". On the other hand, the existence of phase transition in various models has been established by model-dependent methods. It would be valuable to find a unified method for establishing the existence of phase transitions. Of course, it is unrealistic to expect one method to be applicable to all models. However, it would be significant to find a method which is applicable to a fairly wide class of models.

Problem 22. Suppose that an approximately inner dynamics $\{\mathfrak{A}, \rho(t)\}$ with a UHF algebra $\mathfrak{A}$ has no pregenerator δ_0 satisfying the approximate boundedness condition. Then under some minor additional conditions, can we conclude that $\{\rho(t)\}$ has phase transition at some β $(\beta \in (-\infty, \infty))$?

Problem 23. Is there some stability condition at β which includes the approximate boundedness condition and assures the uniqueness of KMS at β ?

Remark. Stability conditions under bounded perturbations, which all KMS states satisfy, have been studied in [2], [29].

Problem 24. Suppose that a C*-dynamics $\{\mathfrak{A}, \rho(t)\}$ has a unique ground state. If $\{\rho(t)\}$ has a KMS state at a sufficiently large β, can we conclude that $\{\rho(t)\}$ has no phase transition at β ?

Remark. In the Ising ferromagnet, this problem has an affirmative answer because of Griffiths's inequality.

For commutative derivations in UHF algebras, we have a fairly detailed description of all KMS states at β $(-\infty < \beta < +\infty)$ ([66],[67],[68]). It is not unreasonable, therefore, to hope for a comparable development of phase transition theory. This is certainly one of the most important aspects of the theory of unbounded derivations in C*-algebras.

§9. Related matters. Let S be a symmetric operator in a Hilbert space $\mathfrak{H}$. In mathematical physics, one is often faced with the problem of extending S to a self-adjoint operator. In many cases, we may assume that there is an increasing sequence $\{V_n\}$ of closed subspaces in $\mathfrak{D}(S)$ such that $\bigcup_{n=1}^{\infty} V_n$ is dense in $\mathfrak{H}$.

Theorem 9.1 ([35]). If $\|SP_n - P_nSP_n\| = O(1)$, then the closure of S is self-adjoint.

Theorem 9.2 ([35]). If for a fixed k, $SV_n \subset V_{n+k}$ $(n = 1, 2, \ldots)$ and $\|SP_n - P_nSP_n\| = O(\lambda_n)$ and $\sum_{n=1}^{\infty} \frac{1}{\lambda_n} = +\infty$, then the closure of S is self-adjoint.

It would be desirable to explore further the self-adjoint extension problem or self-adjoint closure problem along these lines.

Also, I would like to mention articles on other subjects as references. [5], [7], [8], [12], [13], [14], [15], [20], [21], [26], [30], [31], [32], [38], [39], [43], [44], [48], [62] .

Acknowledgement

I express sincere thanks to my colleagues, R. V. Kadison and R. T. Powers, who have read the manuscript and have given me many valuable suggestions.

REFERENCES

1. H. Araki, On the uniqueness of KMS states of one-dimensional quantum lattice system, Comm. Math. Phys. 44(1975), 1-7 .
2. H. Araki and G. L. Swell, KMS conditions and local Thermo-dynamical stability of Quantum lattice system, Comm. Math. Phys., 52(1977), 103-110.
3. W. Arveson, On groups of automorphisms of operator algebras, Jour. Func. Anal. 15(1974), 217-243.
4. H. J. Borchers, Energy and momentum as observables in quantum field theory, Comm. Math. Phys. 2(1966), 49-54.
5. H. J. Brascamp, Equilibrium states for a classical lattice gas, Comm. Math. Phys. 18(1970), 82-96.
6. O. Bratteli, Self-adjointness of unbounded derivations of C*-algebras, preprint.
7. O. Bratteli and U. Haagerup, Unbounded derivations and invariant states, preprint.
8. O. Bratteli, R. Herman and D. W. Robinson, Perturbations of Flows on Banach Spaces and Operator Algegras, preprint.
9. O. Bratteli and D.W. Robinson, Unbounded derivations of C*-algebras, Comm. Math. Phys. 42(1975) 253-268.
10. ______________________________, Unbounded derivations of C*-algebras II, Comm. Math. Phys. 46(1976), 11-30.
11. ______________________________, Unbounded derivations of von Neumann algebras, to appear in Ann. Inst. H. Poincare.
12. ______________________________, Unbounded derivations and invariant trace state, Comm. Math. Phys. 46(1976), 31-35.

13. G. Brink, On a class of approximately inner automorphisms of the C.A.R.-algebra, a preprint.

14. G. Brink and M. Winnink, Spectra of Liouville operators, Comm. Math. Phys. 51(1976),135-150.

15. O. Buchholtz and J.E. Roberts, Bounded perturbations of dynamics, Comm. Math. Phys. 49(1976),161-177.

16. D. P. Chi, Derivations in C*-algebras, Dissertation, University of Pennsylvania.

17. I. Colojoara and C. Foias, Theory of generalized spectral theory, Gordon and Breach, 1968.

18. J. Cuntz, Locally C*-equivalent algebras, Jour. Func. Anal., 23(1976).

19. G. F. Dell'Antonio, On some groups of automorphisms of physical observables, Comm. Math. Phys., 2(1966), 384-397.

20. S. Doplicher, An algebraic spectrum condition, Comm. Math. Phys., 1(1965), 1-5.

21. ____________, A remark on a theorem of Powers and Sakai, Comm. Math. Phys., 45(1975), 59.

22. S. Doplicher, R. V. Kadison, D. Kastler and D.W. Robinson, Asymptotically abelian systems, Comm. Math. Phys., 6(1957), 101-120.

23. S. Doplicher, D. Kastler, and E. Størmer, Invariant states and asymptotic abelianness, Jour. Func. Anal., 3(1969), 21-26.

24. G. Elliott, Derivations of matroid C*-algebras, Inventions Math. 9(1970), 253-269.

25. __________, Some C*-algebras with outer derivations III, preprint.

26. G. Gallavotti and M. Pulvirenti, Classical KMS condition and Tomita-Takesaki theory, Comm. Math. Phys., 46(1976), 1-9.

27. L. Gross, Existence and uniqueness of physical ground states, Jour. Func. Anal., 10(1972), 52-109.

28. R. Haag, N. Hugenholtz and M. Winnink, On the equilibrium states in quantum statistical mechanics, Comm. Math. Phys., 5(1967), 215-236.

29. R. Haag, D. Kastler and E. B. Trych-Pohlmeyer, Stability and equilibrium states, Comm. Math. Phys., 38(1974), 173-193.

30. A. Helemski and Ya. Sinai, A description of differentiations in algebras of the type of local observables of Spin systems, Func. Anal. Appl., 6(1973), 343-344.

31. R. Herman, Unbounded derivations, Jour. Func. Anal., 20(1975), 234-239.

32. _________, Unbounded derivations, to appear.

33. E. Hille and R. Philips, Functional analysis and semi-groups, Amer. Math. Soc. Colloquium publication, Vol. 31, Rhode Island, 1957.

34. B. Johnson and A. Sinclair, Continuity of derivations and a problem of Kaplansky, Amer. J. Math., 90(1968), 1067-1073.

35. P. Jørgensen, Approximately reducing subspaces for unbounded linear operators, Jour. Func. Anal., 23(1976), 392-414.

36. ___________, Trace states and KMS states for approximately inner dynamical one-parameter group of *-automorphisms, to appear in Comm. Math. Phys.

37. ___________, Approximately invariant subspaces for unbounded linear operators, to appear in Math. Ann.

38. ____________, Unbounded derivations in operator algebras and extensions of states, to appear in Tohoku Math. J.

39. ____________, and C. Radin, Approximately inner dynamics, preprint.

40. S. Kantorovitz, Classification of operators by means of their operator calculus, Trans. Amer. Math. Soc., 115(1965), 192-214.

41. T. Kato, Perturbation theory for linear operators, Berlin-Heiderberg-New York, Springer-Verlag, 1966.

42. A. Kishimoto, Dissipations and derivations, Comm. Math. Phys., 47(1976), 25-32.

43. ____________, On uniqueness of KMS states of one-dimensional quantum lattice systems, Comm. Math. Phys., 47(1976), 167-170.

44. ____________, Equilibrium states of a semi-quantum lattice system, preprint.

45. P. Kruszyński, On existence of KMS states for invariantly approximately inner dynamics, Bull. Acad. Pol. Sci.

46. C. Lance and A. Niknam, Unbounded derivations of group C*-algebras, preprint.

47. O. Lanford and D. Ruelle, Integral representations of invariant states on a B*-algebra, J. Math. Phys., 8(1967), 1460-1463.

48. G. Lindblad, On the generators of quantum dynamical semi-groups, Comm. Math. Phys., 48(1976), 147.

49. R. Longo, On perturbed derivations of C*-algebras, preprint.

50. G. Lumer and R. S. Philips, Dissipative operators in a Banach space, Pacific J. Math., 11(1961), 679-698.

51. R. McGovern, Quasi-free derivations on the canonical anti-commutation relation algebra, to appear in J. Func. Anal.

52. A. McIntosh, Functions and derivations of C*-algebras, preprint.

53. B. Sz-Nagy, On uniformly bounded linear transformations in Hilbert space, Acta. Sci. Math. (Szeged) 11(1947), 152-157.

54. E. Nelson, Analytic vectors, Ann. of Math. 70(1959), 572-615.

55. D. Olesen and G. Pedersen, Groups of automorphisms with spectrum condition and the lifting problem, Comm. Math. Phys., 51(1976), 85-95.

56. S. Ôta, Certain operator algebras induced by *-derivations in C*-algebras on an indefinite inner product space, preprint.

57. G. Pedersen, Lifting derivations from quotients of separable C*-algebras, Proc. Nat. Acad. Sci.,USA, 73(1976), 1414-1415.

58. R. T. Powers, Representations of the canonical anti-commutations relations, Thesis, Princeton (1967).

59. R. T. Powers, A remark on the domain of an unbounded derivation of a C*-algebra, Jour. Func. Anal., 18(1975), 85-95.

60. R. T. Powers and S. Sakai, Existence of ground states and KMS states for approximately inner dynamics, Comm. Math. Phys., 39(1975), 273-288.

61. ______________________________, Unbounded derivations in operator algebras, Jour. Func. Anal., 19(1975), 81-95.

62. D. W. Robinson, The approximation of flow, Jour. Func. Anal., 24(1977), 280-290.

63. D. Ruelle, Statistical mechanics: Rigorous results, W. A. Benjamin, New York (1969).

64. S. Sakai, C*-algebras and W*-algebras, Berlin-Heidenberg-New York, Springer-Verlag, 1971.

65. ____________, On one-parameter subgroups of *-automorphisms on operator algebras and the corresponding unbounded derivations Amer. J. Math., 98(1976), 427-440.

66. ____________, On commutative normal *-derivations, Comm. Math. Phys., 43(1975), 39-40.

67. ____________, On commutative normal *-derivations II, Jour. Func. Anal., 21(1976), 203-208.

68. ____________, On commutative normal *-derivations III, Tohoku Math. J., 28(1976), 583-590.

69. J. G. Stamfli, Derivations on $B(\mathcal{H})$: the range, Illinois J. Math., 17(1973), 518-524.

70. H. F. Trotter, Pacific J. Math., 8(1958), 887-919.

71. K. Yoshida, Functional Analysis, Berlin-Heidenberg-New York, Springer-Verlag, 1974 (Fourth edition).

QUASI-EXPECTATIONS AND INJECTIVE OPERATOR ALGEBRAS

John W. Bunce and William L. Paschke

Let M be a von Neumann algebra acting on Hilbert space H. By a <u>quasi-expectation</u> we mean a bounded linear projection $E: B(H) \longrightarrow M$ such that $E(xay) = xE(a)y$ for a in B(H), x and y in M. An old result of J. Tomiyama [6] says that any projection of norm one from B(H) onto M is a quasi-expectation.

<u>Theorem 1</u>. Let M be a countably decomposable finite von Neumann algebra contained unitally in a C*-algebra A. Then the following are equivalent:

a) There exists a continuous linear functional f on A such that $f|_M$ is a faithful normal trace and $f(ax) = f(xa)$ for all x in M and a in A.

b) There is a positive functional g on A such that $g(ax) = g(xa)$ for all x in M, a in A, and $g|_M$ is faithful.

c) There exists a projection of norm one from A to M.

<u>Sketch of proof</u>. a) implies b) is standard and c) implies a) is clear. To prove that b) implies c), note that $\tau = (g|_M)_n$ (the normal part of $g|_M$ as in [3], [4]) is a faithful normal trace on M. Then use Sakai's Radon-Nikodym theorem to show that for each a in A, there is a unique element E(a) in M such that $((g \cdot a)|_M)_n = \tau \cdot E(a)$. Then $E: A \longrightarrow M$ is the desired projection.

We remark that Theorem 1 is proved when $A = B(H)$ (although not explicitly stated) by Connes in [1], using the concept of semidiscreteness [2].

<u>Theorem 2</u>. Let M be a von Neumann algebra acting on H. If there exists a quasi-expectation of B(H) onto M, then there exists a projection of norm one of B(H) onto M (that is, M is injective).

Theorem 2 is proved in the finite countably decomposable case by using Theorem 1. Standard techniques then prove the semifinite case, and the decomposition of a properly infinite von Neumann algebra as a crossed product of a semifinite algebra by a one-parameter group of automorphisms is used to prove the general case.

Recall that a von Neumann algebra M is said to be amenable as a von Neumann algebra if every normal derivation of M into a dual normal M-module is inner.

Theorem 3. If M is an amenable von Neumann algebra acting on H, then there exists a quasi-expectaion of B(H) onto M'.

Hence if M is amenable, then M' is injective by Theorem 2, so M is injective by Tomita's theorem. So we have

Corollary. (Connes) If M is an amenable von Neumann algebra, then M is injective. If A is an amenable C*-algebra, then A is nuclear.

Sketch of proof of Theorem 3. Make $Y = B(B(H), B(H))$ into a dual normal M-bimodule by defining $(G\cdot x)(b) = G(b)x$ and $(x\cdot G)(b) = xG(b)$ for x in M, G in Y, and b in B(H). The predual of Y is $B(H)\hat{\otimes}T(H)$, where T(H) is the trace class operators. Let $W = \{G \text{ in } Y: \langle G, bs\otimes t - b\otimes st\rangle = 0, \langle G, sb\otimes t - b\otimes ts\rangle = 0, \langle G, s\otimes t\rangle = 0$ for b in B(H), t in T(H), s in $M'\}$. Then W is a dual normal M-bimodule. Let F in Y be defined by $F(b) = b$, and let $\delta: M \longrightarrow Y$ be the derivation given by $\delta(x) = F\cdot x - x\cdot F$. Note that $\delta(M)$ is contained in W, so $\delta: M \longrightarrow W$. By the amenability of M, there exists a G in W with $\delta(x) = G\cdot x - x\cdot G$ for all x in M. Then $E: B(H) \longrightarrow B(H)$ given by $E(b) = b - G(b)$ is a quasi-expectation of B(H) onto M'.

References

1. A. Connes, On the cohomology of operator algebras, preprint.
2. E. G. Effros and E. C. Lance, Tensor products of operator algebras, to appear in Adv. Math.
3. M. Takesaki, On the conjugate space of an operator algebra, Tohoku Math. J. 10 (1958), 194-203.
4. M. Takesaki, On the singularity of a positive linear functional on operator algebra, Proc. Japan Acad. 35 (1959), 365-366.
5. M. Takesaki, Duality in crossed products and the structure of von Neumann algebras of type III, Acta. Math. 131 (1973), 249-310.
6. J. Tomiyama, On the projection of norm one in W*-algebras, Proc. Japan Acad. 33 (1957), 608-612.

GENERAL SHORT EXACT SEQUENCE THEOREM FOR TOEPLITZ OPERATORS OF UNIFORM ALGEBRAS

Jun TOMIYAMA and Kôzô YABUTA
Faculty of Science, Yamagata University
Yamagata, Japan
and
College of Technology
Kyôto Technical University
Kyôto, Japan

We recall first the classical case of Toeplitz operators. Let $C(T)$ be the algebra of all complex valued continuous functions on the unit circle T in the complex plane and A the disk algebra. Let $H^2(T)$ be the Hardy space in $L^2(T)$ with respect to the normalized Lebesgue measure m. Let p be the orthogonal projection of $L^2(T)$ onto $H^2(T)$. A Toeplitz operator T_ϕ with bounded measurable symbol $\phi \in L^\infty(T)$ is defined as $T_\phi(f) = P(\phi f)$ for $f \in H^2(T)$. Denote by $\mathcal{J}(T,A)$ (resp. $\mathcal{J}(T,H^\infty(T))$ the C*-algebra generated by the set $\{T_\phi : \phi \in A\}$ (resp. $\{T_\phi : \phi \in H^\infty(T)\}$). Let $\mathcal{C}(T,A)$ (resp $\mathcal{C}(T,H^\infty(T))$ be the commutator ideal of $\mathcal{J}(T,A)$ (resp. $\mathcal{J}(T,H^\infty(T))$. Then it is known that there exists a $*$-homomorphism ρ of $\mathcal{J}(T,A)$ onto $C(T)$ such that the following short sequence

$$\{0\} \longrightarrow \mathcal{C}(T,A) \xrightarrow{\ i\ } \mathcal{J}(T,A) \xrightarrow{\ \rho\ } C(T) \longrightarrow (0)$$

is exact and $\rho(T_\phi) = \phi$ where i is the inclusion map. Further in this case, $\mathcal{C}(T,A)$ coincides with the ideal $\mathcal{LC}(H^2(T))$ of compact operators in $H^2(T)$, so that we get an exact sequence

$$\{0\} \longrightarrow \mathcal{LC}(H^2(T)) \xrightarrow{i} \mathcal{J}(T,A) \xrightarrow{\rho} C(T) \longrightarrow (0).$$

n the other hand, for the C*-algebra $\mathcal{J}(T,H^\infty(T))$ we also get a short xact sequence

$$\{0\} \longrightarrow \mathcal{C}(T,H^\infty(T)) \xrightarrow{i} \mathcal{J}(T,H^\infty(T)) \xrightarrow{\rho} L^\infty(T) \longrightarrow (0)$$

ith $\rho(T_\phi) = \phi$. These results have been extended to many cases, to ther domains in C or C^n [1], [3], [6], [11], and in Stein spaces quite recently)[8]. Moreover there are other short exact sequence theo-ems as in the case of Toeplitz operators (Wiener-Hopf operators) with lmost periodic symbols. In many of these cases, so far as the commutator deals are concerned, the proofs make use of the elegant theorem of Bunce 2] on the joint approximate point spectrum of a commuting family of hypo-ormal operators

Here we propose to prove, in a rather abstract setting, a short xact sequence theorem for Toeplitz operators of a uniform algebra which s general enough to include all of the previous results. It should be oted, however, that whether or not the commutator ideal of the algebra oincides with the algebra of compact operators is another problem. Our roof does not use Bunce's theorem but it is effected by modifying his dea in [2].

Let $C(X)$ be the algebra of all complex valued continuous func-ions on a compact space X and A be a uniform algebra on X. We con-ider a linear representation τ of $C(X)$ into the algebra $L(H)$ of ll bounded linear operators on a Hilbert space H. Assume that atisfies the following conditions:

(1) τ is contractive and $\tau(1) = 1$, the identity operator.

(2) τ is isometric on the algebra A.

(3) $\tau(\phi)\tau(\varphi) = \tau(\phi\varphi)$ for all $\phi \in C(X)$ and $\varphi \in A$.

et $\mathcal{J}(X,A)$ be the C*-algebra generated by the set $\{\tau(\varphi) : \varphi \in A\}$, r $\{\tau(\phi) : \phi \in C(X)\}$, and let Δ be the space of characters of $\mathcal{J}(X,A)$. Then, by the condition (3), Δ can be embedded into X if e consider the map $\beta \in \Delta \longrightarrow (\beta|\ \tau(C(X))) \circ \tau$, a character of $C(X)$. enote by $\Gamma(\tau)$ the image of Δ in X. Our result is the following:

Theorem. $\Gamma(\tau)$ is a closed boundary for A. Moreover, there s a $*$-homomorphism ρ of $\mathcal{J}(X,A)$ onto $C(\Gamma(\tau))$ such that the short equence

$$(0) \longrightarrow \mathcal{C}(X,A) \xrightarrow{i} \mathcal{J}(X,A) \xrightarrow{\rho} C(\Gamma(\tau)) \longrightarrow (0)$$

is exact and $\rho(\tau(\phi)) = \phi|\Gamma(\tau)$ for every $\phi \in C(X)$.

In most of the examples the space X coincides with the Shilov boundary $\Gamma(A)$ of A, so that we get an exact sequence

$$(0) \longrightarrow \mathcal{C}(X,A) \xrightarrow{i} \mathcal{J}(X,A) \xrightarrow{\rho} C(X) \longrightarrow (0)$$

and the isometry $\|\tau(\phi)\| = \|\phi\|$ for every $\phi \in C(X)$. The theorem can also be applied to the representation τ of the couple $(L^\infty(\mu), H^\infty(\mu))$ for a finite nonnegative regular Borel measure μ on X provided that $H^\infty(\mu)$ separates the characters of $L^\infty(\mu)$. Thus, if $\Gamma(H^\infty(\mu))$ is shown to be equal to the maximal ideal space of $L^\infty(\mu)$, $(L^\infty(\mu))$ we get an exact sequence

$$(0) \longrightarrow \mathcal{C}(X,H^\infty(\mu)) \xrightarrow{i} \mathcal{J}(X,H^\infty(\mu)) \xrightarrow{\rho} L^\infty(\mu) \longrightarrow (0)$$

with $\rho(\tau(\phi)) = \phi$ for every $\phi \in L^\infty(\mu)$.

In a setting of Toeplitz operators, the representation τ arises usually as the compression of the multiplication operators to the (abstract) Hardy space $H^2(\mu)$ and the assumptions (1) and (3) are easily see- to hold in this case. In this general setting, condition (2) is not difficult to verify. It is a consequence of the following computation:

$$\left(\int |\varphi|^j d\mu\right)^{1/j} = \left(\int |T_\varphi^j 1|\, d\mu\right)^{1/j} < \|T_\varphi^j 1\|_2^{1/j} < \|T_\varphi\| \ \|1\|_2^{2/j}$$

for every $\varphi \in H^\infty(\mu)$ and $J = 1, 2, 3, \dots$. Letting $J \longrightarrow \infty$ we have $\|\varphi\|_{L^\infty(\mu)} \le \|T_\varphi\|$, which shows that $\|\varphi\|_{L^\infty(\mu)} = \|T_\varphi\|$. Therefore, if the support of μ contains $\Gamma(A)$, the map $\varphi \longrightarrow T_\varphi$ is also an isometry on the algebra A. The computation is effective even when μ is not a bounded measure. In fact, we have for every $\varphi \in H^\infty(\mu)$ and $f \in H^2(\mu)$

$$\int |\varphi^j f|^2 d\mu = \|T_\varphi^j f\|_2^2 \le \|T_\varphi\|^{2j} \ \|f\|_2^2$$

and

$$\left(\int |\varphi^j f|^2 d\mu\right)^{1/2j} \le \|T_\varphi\| \ \|f\|_2^{1/j}$$

Letting $j \longrightarrow \infty$, we see that $\|\varphi\|_{L^\infty(|f|^2 d\mu)} < \|T_\varphi\|$, which implies the inequality $\|\varphi\|_{L^\infty(\mu)} \le \|T_\varphi\|$ as the case may be. This covers the case of Toeplitz operators with almost periodic symbls.

Now let $Q(A)$ be the Choquet boundary for A. We employ the erminology (states) and (pure states) for A in referring to those functionals α with $\alpha(1) = 1 = \|\alpha\|$ and those extreme points in the set f states of A. It is known that a point t of $Q(A)$ corresponds to pure state of A, where it gives rise to a pure state α_t of $\tau(A)$. he following is a key lemma in the proof of the theorem.

Lemma. The state α_t extends uniquely to a pure state of $\mathcal{J}(X,A)$ nd the extended state is a character of $\mathcal{J}(X,A)$.

Sketch of the proof. It suffices to show that a state extension of α_t is necessarily a character of $\mathcal{J}(X,A)$. Note first that $(\tau(\phi)) = \phi(t)$ for every $\phi \in C(X)$ by the uniqueness of the state exensions of the state $t|A$ to $C(X)$. Hence, from the third condition or τ one sees that the left kernel of α contains $\tau(\varphi) - \varphi(t)$ for very A. Therefore,

$$\hat{\alpha}(S(\tau(\varphi) - \varphi(t))) = 0 \quad \text{for every operator } S \text{ in } \mathcal{J}(X,A),$$

hich implies that

$$\hat{\alpha}(ST) = \hat{\alpha}(S)\,\hat{\alpha}(T) \quad \text{for every } S \in \mathcal{J}(X,A) \text{ and } T \in \tau(A).$$

n the other hand, as τ is a completely positive map it has a special orm by the theorem of Stinespring [9]. From this one can verify that very operator T of $\tau(A)$ is hyponormal, i.e. $TT^* \leq T^*T$ (or even subormal). This converts the role of the operator S in the above identity $\hat{\alpha}(ST) = \hat{\alpha}(S)\,\hat{\alpha}(T)$ and tells us that $\hat{\alpha}$ is really multiplicative on (X,A).

In conclusion we remark that arguments for general Toeplitz operators for the uniform algebra A naturally follow after the theorem aong the same lines as the corresponding results in the standard literture. We leave the details to our paper [10].

REFERENCES

[1] M. B. Abrahamse, Toeplitz operators on multiply-connected regions, Amer. J. Math., 96(1974), 261-297.

[2] J. Bunce, The joint spectrum of commuting nonnormal operators, Proc. Amer. Math. Soc., 29(1971), 499-505.

[3] L. A. Coburn, Toeplitz operators on odd spheres, Springer Lecture Notes, 345(1973), 7-12.

[4] L. A. Coburn and R. G. Douglas, C*-algebras of operators on a half space I, IMES Publ. Math., 40(1971), 59-67.

[5] R. G. Douglas, Banach algebra techniques in the theory of Toeplitz operators, CBMS 15, Amer. Math. Soc., Providence, 1973.

[6] I. Janas, Toeplitz operators related to certain domains in C^n, Studia Math., 54(1975), 73-79.

[7] J. Janas, Toeplitz operators for a certain class of function algebras, Studia Math., 55(1975), 157-161.

[8] H. Sato and K. Yabuta, Toeplitz operators on strongly pseudo-convex domains in Stein spaces, preprint.

[9] W. F. Stinespring, Positive functions on C*-algebras, Proc. Amer. Math. Soc., 6(1955), 211-216.

[10] J. Tomiyama and K. Yabuta, Toeplitz operators for uniform algebras to appear in Tôhoku Math. J.

[11] U. Venugopalkrishna, Fredholm operators associated with strongly pseudo-convex domains in C^n, J. Functional Analysis, 9(1972), 349-372.

[12] K. Yabuta, A remark to a paper of Janas; Toeplitz operators related to a certain domains in C^n, to appear.

[13] W. Żelazko, On a problem concerning joint approximate point spectra, Studia Math., 45(1973), 239-240.

AW*-ALGEBRAS WITH MONOTONE CONVERGENCE PROPERTY AND TYPE III, NON W*, AW*-FACTORS

Kazuyuki SAITÔ
Mathematical Institute
Tôhoku University
Sendai, Japan

In 1951, I. Kaplansky [3] introduced a class of C*-algebras called .W*-algebras to separate the discussion of the internal structure of a '*-algebra from the action of its elements on a Hilbert space and showed .hat much of the "non-spatial theory" of W*-algebras can be extended to .W*-algebras.

Every W*-algebra is, of course , AW*, but the converse is not true ıs is shown by Dixmier [1] with an abelian example. I. Kaplansky [4] ›roved that an AW*-algebra of type 1 is a W*-algebra if and only if its :enter is a W*-algebra and conjectured that an AW*-algebra is a W*-algebra .f and only if its center is a W*-algebra.

In 1970, O. Takenouchi [6] and Dyer [2], independently, showed :his to be false by counter examples (non W*, AW*-factors). In 1976, J.). Maitland Wright [8] defined "regular σ-completion" of a separable C*-ılgebra and proved that the regular σ-completion of an infinite dimen-:ional simple separable C*-algebra is a type III, non W*, σ-finite AW*-factors.

In this talk, I would like to present a modification of J. D. Maitland Wright's theorem and, using this, show that the non W*, AW*-factors given by Takenouchi and Dyer are also type III AW-factors.

1. <u>An AW*-algebra with the monotone convergence property (M. C. P.).</u> Let M be an AW*-algebra with the M.C.P. in the sense that every bounded increasing sequence $\{x_n\}$ of self adjoint elements has a least

upper bound $\operatorname{Sup} x_n$ in M.

Then one can easily check that for any increasing sequence of projections $\{p_n\}$ in M, $\operatorname{Sup}_n p_n$ is a projection. Keeping this in mind, we have

Theorem. Let M be an AW*-factor with M.C.P.. Suppose that M has a faithful state (normal or not normal) and is semi-finite, then M is a σ-finite W*-algebra. The assumption of semi-finiteness cannot be dropped.

The key point of the proof is to construct sufficiently many c.a. states on M (c.a. state means that the state is completely additive on projections).

Remark. J. D. Maitland Wright [7] proved, without the assumption of M.C.P., but under the condition that M is finite, that the above statement holds.

2. AW*-factors constructed by Takenouchi and Dyer.

Let Z be an abelian AW*-algebra and let G be an abelian group of $*$-automorphisms of Z with an action $a \longrightarrow a^g (a \in Z,\ g \in G)$. One can construct a faithful AW*-module $\mathfrak{M}$ over Z ([5]) as $l^2(G,Z) =$ $\{x = (x_g),\ x_g \in Z$ for each $g \in G$, $\Sigma x_g^* x_g$ is in Z (order convergence in Z)$\}$. We know that the set $\mathfrak{B}(\mathfrak{M})$ of all bounded module endomorphisms ("operators") of $\mathfrak{M}$ is a type 1 AW*-algebra with center Z. Define two types of "operators" on $\mathfrak{M}$ as follows:

$$\left.\begin{array}{ll} La: \{x_g\} \longrightarrow \{a^g x_g\} & \\ U_h: \{x_g\} \longrightarrow \{y_g\} \quad \text{where } y_g = x_{g-h} & \end{array}\right\} \begin{array}{l} \{x_g\} \in \mathfrak{M} \\ a \in Z,\ h \in G. \end{array}$$

Let $\mathbb{M}(Z,G)$ be the smallest "weakly closed" AW*-subalgebra of $\mathfrak{B}(\mathfrak{M})$ containing $\{La, U_h;\ a \in Z,\ h \in G\}$. Takenouchi showed that, under the condition that the action of G on Z is free and ergodic, $\mathbb{M}(Z,G)$ is an AW*-factor such that $\{La;\ a \in Z \equiv \tilde{Z}\}$ is a maximal abelian $*$-subalgebra and gave an example of (Z,G) as follows:
Let Z be the algebra of all bounded Baire functions on $[0, 1)$ modulo the sets of 1st Category and let G_θ be the group of $*$-automorphisms of Z which are induced naturally by the group of homeomorphisms $\{\theta m + n;\ n, m = 0, \pm 1, \pm 2, \ldots\}$ (mod 1) of $[0, 1)$ (where θ is an irrational number). Then $\mathbb{M}(Z,G)$ is a non-W*, AW*-factor.

Dyer's example is the following: Let H be a Hilbert space with an orthogonal basis $\{e_x;\ 0 \leq x < 1,\ x$ a real number$\}$. Any operator A on H has a matrix representation $A_{x,y} = (Ae_y, e_x)$ where $0 \leq x,$

< 1. Let $\mathfrak{A}$ (resp. $\mathfrak{J}$) be the set of operators A on H with mat-
ices $A_{x,y}$ with

(1) $A_{x,y} = 0$ except when $y - x = j \cdot 2^{-k}$ for some integer

$k \geqq 1$, and $-2^k < j < 2^k$ (integer)

(2) for $k \geqq 1$, and $0 \leqq i$, $j < 2^k$, the function defined

or $x \in [0, 1)$ by $f(x) = A_{2^{-k}(i + x), 2^{-k}(j + x)}$ is a bounded Baire

unction (resp. $\{x; 0 \leqq x < 1, f(x) \neq 0\}$ is contained in a set of Ist ategory in $[0, 1)$).

Then $\mathfrak{A}$ is a C*-algebra with a closed two sided ideal $\mathfrak{J}$ and $\mathfrak{A}/\mathfrak{J}$ is a non W*, AW*-factor with a maximal abelian $*$-subalgebra which s $*$-isomorphic with the above Z.

3. Types of the above AW*-factors. Keeping the notations of §2, et G_0 be the group of $*$-automorphisms of Z naturally induced by the roup of homeomorphisms corresponding to the dyadic rationals of [0, 1). . D. Maitland Wright [8] tells us that Z has a faithful state ψ. by he construction of $\mathbb{M}(Z,G_\theta)$, one can easily check that there is a aithful positive projection Φ of $\mathbb{M}(Z,G_\theta)$ onto $\mathcal{Z}$. Let $\phi = \psi \circ \Phi$. t is easily proved that ϕ is a faithful state on $\mathbb{M}(Z,G_\theta)$. Note that (Z,G_θ) has the "M.C.P.", by Theorem of 1, if $\mathbb{M}(Z,G_\theta)$ is semi-finite, hen $\mathbb{M}(Z,G_\theta)$ is a W*-algebra, but this is a contradiction and $\mathbb{M}(Z,G_\theta)$ s of Type III. A straightforward verification tells us that also $\mathfrak{A}/\mathfrak{J}$ as the "M.C.P." and has a maximal abelian $*$-subalgebra which is $*$-iso- orphic to Z and, onto which $\mathfrak{A}/\mathfrak{J}$ has a faithful positive projec- ion map. Thus, by the same reasoning, $\mathfrak{A}/\mathfrak{J}$ is of Type III. Moreover, $\mathfrak{A}/\mathfrak{J}$ is $*$-isomorphic with $\mathbb{M}(Z,G_0)$. Remark that both $\mathbb{M}(Z,G_\theta)$ and (Z,G_0) are σ-finite, and do not have any non-trivial separable repre- entations.

The following question remains open:

Are $\mathbb{M}(Z,G_\theta)$ and $\mathbb{M}(Z,G_0)$ $*$-isomorphic ?

REFERENCES

[1] J. Dixmier, Sur certains espaces considérés par M. H. Stone, Summa Brasil. Math., 2(1951), 151-182.

[2] J. Dyer, Concerning AW*-algebras, To appear in J. Functional Analysis.

[3] I. Kaplansky, Projections in Banach algebras, Ann. of Math., 53 (1951), 235-249.

[4] I. Kaplansky, Algebras of Type 1, Ann. of Math., 56(1952), 460-470.

[5] I. Kaplansky, Modules over operator algebras, Amer. J. Math., 75(1953), 839-858.

[6] O. Takenouchi, Note following this.

[7] J. D. Maitland Wright, On AW*-algebras of finite type, J. London Math. Soc., 12(1976), 431-439.

[8] J. D. Maitland Wright, Wild AW*-factors and Kaplansky-Rickart algebras, J. London Math. Soc., 13(1976), 83-89.

A non-W*, AW*-factor

Osamu Takenouchi

The aim of this report is to exhibit an example of an AW*-algebra which is not a W*-algebra. This example will be constructed by a crossed product method in AW*-algebras.

1. Construction of an AW*-module

Let $\underline{A}$ be a commutative AW*-algebra and G a group of *-automorphisms of $\underline{A}$. We construct an AW*-module H of I. Kaplansky using G as the set of indices and $\underline{A}$ as the coefficient domain. (cf. [1]) Then the set $\underline{B}$ of bounded operators appears to be an AW*-algebra by virtue of Kaplansky's work. Here we mean by a bounded operator a bounded linear operator which, at the same time, is a homomorphism with respect to the $\underline{A}$-module structure. To each bounded operator A is associated a matricial expression

$$A \sim \langle a_{g,k} \rangle_{g,k \in G}$$

This means, when $\{x_g\}$ is an element of H whose components are all zero except for the index k then

$$A\{x_g\} = \{y_g\}, \quad \text{where} \quad y_g = a_{g,k}x_k$$

Hereafter, u_k is used to denote an element of H which have all its components zero except for the index k where it has a component 1.

2. Construction of an AW*-algebra

Let $\underline{M}$ be the set of those bounded operators on H which have a matricial expression of the following form :

$$A \sim < a_{g,k} > \quad , \quad a_{g,k} = (a_{gk^{-1},e})^k \quad \text{for } g, k \in G .$$

$\underline{M}$ is a sub-AW*-algebra of $\underline{B}$.

$\underline{M}$ is the AW*-algebra generated by the following types of operators L_a and U_h :

$$L_a : \{x_g\} \to \{a^g \cdot x_g\}$$

$$U_h : \{x_g\} \to \{y_g\} , \quad \text{where } y_g = x_{h^{-1}g}$$

Proof of the fact that $\underline{M}$ is a sub-AW*-algebra of $\underline{B}$.

According to the definition of a sub-AW*-algebra, what we must show is that, for each subset $\underline{S}$ of $\underline{M}$, the left annihilator $\underline{N}$ of $\underline{S}$ in $\underline{B}$ is generated by a projection E of $\underline{M}$.

Now a left annihilator of $\underline{S}$ is an annihilator of the image under S . Let E_0 be the projection on the sub-AW*-module M of H generated by the image under $\underline{S}$. Then we have $E = I - E_0$. So we will see that $E_0 \in \underline{M}$.

For this aim we introduce an operator $V_h (h \in G)$.

$$V_h : \{x_g\} \to \{y_g\} ,$$

if $\{y_g\}$ defined by

$$y_g = (x_{gh})^{h^{-1}}$$

is in H . This is not a homomorphism of module. But one sees that, for any $A \in \underline{M}$, Au_k is in the domain of V_h and

$$V_h A u_k = A V_h u_k = A u_{kh^{-1}}$$

(u_k is the element introduced at the end of section 1.)

A remarkable thing is that the sub-AW*-module M is contained in the domain of V_h and invariant under V_h. In fact, let M_0 be the set of elements of the following form

$$\sum_{\text{finite sum}} (a_j)^{h_j} A_j u_{h_j}, \quad \text{where } h_j \in G, \quad a_j \in \underline{A}, \quad A_j \in \underline{S}.$$

Applying V_h for these elements, they rest in M_0. M is then obtained from M_0 by the following two stages. First, add all the elements of the form $\sum p_\lambda s_\lambda$, where p_λ's are orthogonal projections in $\underline{A}$ summed up altogether to 1, and x_λ's $\in M_0$ are bounded in norm. Then, complete by the norm. By a simple calculus, we can check that, in each stage, the resulting set is stable under V_h. The stability of M under V_h can be written $V_h E_0 = E_0 V_h E_0$. We observe here $E_0 V_h = V_h E_0$. In fact, as

$$(V_h x, y) = \sum (x_{gh})^{h^{-1}} y_g{}^* = (\sum x_g (y_{gh^{-1}}{}^*)^h)^{h^{-1}}$$

$$= (x, V_{h^{-1}} y)^{h^{-1}},$$

we have

$$(E_0 V_h x, y) = (V_h x, E_0 y) = (x, V_{h^{-1}} E_0 y)^{h^{-1}}$$

$$= (V_h E_0 x, E_0 y) = (V_h E_0 x, y)$$

Therefore, for

$$E_0 \sim \langle a_{g,k} \rangle$$

we have

$$E_0 u_k = E_0 V_{k^{-1}} u_e = V_{k^{-1}} E_0 u_e = V_{k^{-1}} \{a_{g,e}\} = \{(a_{gk^{-1},e})^k\}$$

This shows that $E_0 \in \underline{M}$.

3. Some properties on the group

We consider the following properties of freeness and ergodicity.

(F) Taking a non-zero projection p of $\underline{A}$, no element of G different from the identity will leave fixed all the projections smaller than p .

(E) A projection of $\underline{A}$ which is invariant under the action of the elements of G must be 0 or 1 .

4. Lemma. — If (E) is satisfied, the set of L_a $(a \in A)$ forms a maximal abelian subalgebra of $\underline{M}$.

Proof. Let $A \sim < a_{g,k} > \in \underline{M}$, which commutes with any L_a . Then we have

$$a^g \cdot a_{g,k} = a^k \cdot a_{g,k} , \quad \text{for any } a \in \underline{A}$$

Replacing $a_{g,k}$ by its multiple p which is a projection, and suitably choosing a we will have

$$p^h \cdot p = p \quad \text{for any } h \in G .$$

We conclude from this $p = 0$ or 1 under (E), and $A = L_{a_e}$.

5. Lemma — If (F) and (E) are satisfied, the center of $\underline{M}$ is reduced to the scalar multiples of identity and $\underline{M}$ is an AW*-factor.

Proof. An element L_a commutes with U_h if and only if $a^h = a$. If this is satisfied for any $h \in G$, then under (E), a must be a scalar multiple of the identity.

6. A non-W*, AW*-factor

Let $\underline{A}$ be a commutative AW*-algebra which is not a W*-algebra. As such an example, we can take the set of bounded Baire functions on the interval [0, 1] where two functions are to be looked on as equal if they differ only on a set of first category. Here we turn this into an AW*-algebra by taking the addition, multiplication and adjunction pointwise in [0, 1].

Next, G is a group of automorphismes of $\underline{A}$ having the properties (F) and (E). For the particular $\underline{A}$ just mentioned, a traslation on the interval [0, 1] by integral multiples of a fixed irrational number α will give such an example.

The AW*-algebra $\underline{M}$ based on $\underline{A}$ and G, and constructed as in section 2 will then give an example of a non-W*, AW*-factor. It is indeed an AW*-factor as we have observed in section 5, but it is not a W*-algebra. This can be seen from the fact that a maximal abelian subalgebra in a W*-algebra must be a W*-algebra, whereas the set of L_a's in section 4, being isomorphic to $\underline{A}$, is not a W*-algebra.

Reference

[1] I. Kaplansky : Modules over operator algebras, Amer. J. Math. 75 (1953), 839-858.

Fixed points and commutation theorems

A. Van Daele (*)

Let α be a continuous action of a locally compact group G on a von Neumann algebra M. We will prove a theorem which can be used in a number of situations to obtain the fixed point algebra $\{x \in M | \alpha_t(x) = x \text{ for all } t \in G\}$. We will illustrate this in a few special cases and show how certain important commutation theorems follow easily.

Throughout we fix a left invariant Haar measure dt on G, and we work with integrals of σ-weakly continuous M-valued functions on G in the weak sense.

<u>Theorem</u>. Let M be a von Neumann algebra and α a continuous action of G on M. Suppose M_o is a σ-weakly dense *-subalgebra of M such that

i) $\alpha_t(M_o) = M_o$ for all $t \in G$,

ii) for all $a,b \in M_o$ there is a compact subset K of G such that $\alpha_t(a)b = 0$ for t outside K,

iii) $\int \alpha_t(a)dt$ exists for all $a \in M_o$.

Then the fixed point algebra is generated by $\{\int \alpha_t(a)dt | a \in M_o\}$.

<u>Proof</u>. Let R be the von Neumann algebra generated by the elements $\{\int \alpha_t(a)dt | a \in M_o\}$. Clearly any element in R will be a fixed point and therefore it is sufficient to show that any fixed point is in R.

(*) Katholieke universiteit Leuven, Departement Wiskunde, Celestijnenlaan 200B, B-3030 HEVERLEE (Belgium)

Take $a,b \in M_o$, then it can be shown that $\int \alpha_t(axb)dt$ exists for all $x \in M$ and that $x \to \int \alpha_t(axb)dt$ is σ-weakly continuous. If $x \in M_o$ then $axb \in M_o$ and $\int \alpha_t(axb)dt \in R$. Then by continuity this will also be true for all $x \in M$.

Now take any fixed point $x_o \in M$ and define $x(s)$ for any $s \in G$ by

$$x(s) = \int \alpha_t(\alpha_s(a)x_o b)dt.$$

Because also $\alpha_s(a) \in M_o$ we get that $x(s) \in R$ for any s. Let $p \in M_o$, then $\alpha_t(b)p$ has compact support in t and it follows that $x(s)p$ will be continuous in s. Now denote $a_o = \int \alpha_s(a)ds$ and $b_o = \int \alpha_t(b)dt$. Then for any compact K we have that $\int_K \alpha_s(a)ds \to a_o$ as K increases and therefore

$$\int_K x(s)p\,ds = \int \alpha_t(\int_K \alpha_s(a)ds)x_o b)p\,dt \to \int \alpha_t(a_o x_o b)p\,dt = a_o x_o b_o p$$

where we used the fact that the integral is continuous as we are integrating t only over a compact set.

Similarly for any $q \in M_o$ we get

$$\int_K qx(s)ds = \int_K (\int q\alpha_{ts}(a)x_o\alpha_t(b)dt)ds$$

$$= \int_{K^{-1}} (\int q\alpha_t(a)x_o\alpha_{ts}(b)dt)ds$$

$$\to \int q\alpha_t(a)x_o b_o dt = qa_o x_o b_o \ .$$

Now for any $y \in R'$ we have $x(s)y = yx(s)$ and so $qx(s)yp = qyx(s)p$ and by integrating s over K and taking the limit $K \to G$ we find $qa_o x_o b_o yp = qya_o x_o b_o p$. This holds for all $p,q \in M_o$ and therefore $a_o x_o b_o y = ya_o x_o b_o$. Because $a_o, b_o \in R$ and $y \in R'$ also $a_o x_o yb_o = a_o yx_o b_o$.

If we can show that the set $\{b_o\mathcal{H} | b \in M_o\}$ is dense in $\mathcal{H}$ where $\mathcal{H}$ is the Hilbert space on which M acts then it will follow that $a_o x_o y = a_o yx_o$, and similarly because M_o is self-adjoint that $x_o y = yx_o$. Then $x_o \in R'' = R$ and the proof will be complete. So suppose $\xi \perp b_o\mathcal{H}$ for all $b \in M_o$. Then $\int\langle\alpha_s(b)\xi,\xi\rangle = \langle b_o\xi,\xi\rangle = 0$, and with b replaced by b^*b we get $\int\langle\alpha_s(b^*b)\xi,\xi\rangle ds = 0$. Then $\langle\alpha_s(b^*b)\xi,\xi\rangle = 0$ for all s, in particular $\langle b^*b\xi,\xi\rangle = 0$ and hence $b\xi = 0$. Therefore $\xi = 0$ and the proof is complete. (For a similar proof see section 3 of [7])

Applications. Let us first fix some notations. By $K(G)$ we denote the set of complex-valued continuous functions with compact support on G. The left and right regular representations on $L_2(G)$ are defined by $(\lambda_s f)(t) = f(s^{-1}t)$ and $(\rho_s f)(t) = \Delta(s)^{1/2} f(ts)$ for $f \in K(G)$ and $s,t \in G$. For any bounded continuous complex-valued function g on G we also consider the multiplication operator m_g defined by $(m_g f)(s) = g(s)f(s)$ where $f \in K(G)$ and $s \in G$.

1. Our first application is in the theory of crossed products. Let M be a von Neumann algebra acting on a Hilbert space $\mathcal{H}$, and α a continuous action of G on M. Then we consider $\tilde{M} = M \otimes \mathcal{B}(L_2(G))$ as acting on $\mathcal{H} \otimes L_2(G)$ and we define an action θ of G on $\tilde{M}$ by $\theta_t = \alpha_t \otimes \mathrm{ad}\,\rho_t$. We will apply the theorem to the action θ of G on $\tilde{M}$. For M_o we take the algebra of linear combinations of operators of the form $x \otimes \lambda_s m_f$ where $x \in M$, $s \in G$ and $f \in K(G)$. Then $\theta_t(x \otimes \lambda_s m_f) = \alpha_t(x) \otimes \lambda_s \rho_t m_f \rho_t^* = \alpha_t(x) \otimes \lambda_s m_{f_t}$ where $f_t(r) = f(rt)$ for all $r \in G$, and it follows easily that M_o satisfies all the conditions of the theorem.

   A simple calculation shows that $\int\theta_t(x \otimes m_f)dt = \int f(t)\pi_\alpha(\alpha_t(x))dt$ where $(\pi_\alpha(x)\xi)(s) = \alpha_{s^{-1}}(x)\xi(s)$ for any $\xi \in K(G,\mathcal{H})$, the continuous $\mathcal{H}$-valued functions with compact support on G, considered as a subset of $\mathcal{H} \otimes L_2(G)$ in the usual way. Then a straightforward application of the theorem yields that the fixed point algebra is precisely the von Neumann algebra generated by $\{\pi_\alpha(x), 1 \otimes \lambda_s \mid x \in M, s \in G\}$, i.e. the crossed product $M \otimes_\alpha G$ of M by the action α of G. Now it is well known and easy to show that from this the so-called commutation theorem for crossed products follows (see [1,2,6,7]).

2. Our second application is concerned with the generalized commutation relations of Takesaki [5]. First let $M = \{m_f | f \in K(G)\}''$ as acting on $L_2(G)$ and let $\alpha_t = \mathrm{ad}\, \rho_t$. Take a closed subgroup H of G and restrict the action α to H. We will apply our theorem to the action α of H on M and we take $M_o = \{m_f | f \in K(G)\}$. It is again easy to show that the theorem can be applied and it follows that the fixed point algebra is generated by $\{m_{\tilde f} | f \in K(G)\}$ where $\tilde f(s) = \int_H f(sh)dh$. It is well known that $\tilde f$ can be considered a function in $K(G/H)$, the complex-valued continuous functions with compact support on the left cosets G/H. Moreover every such function is of this type. So we get

$$\{m_f | f \in K(G)\}'' \cap \{\rho_t | t \in H\}' = \{m_f | f \in K(G/H)\}''$$

or equivalently, because $\{m_f | f \in K(G)\}'' = \{m_f | f \in K(G)\}'$

$$\{m_f, \rho_t | f \in K(G), t \in H\}' = \{m_f | f \in K(G/H)\}''.$$

Next take $M = \{m_f, \rho_t | f \in K(G), t \in H\}''$ and consider another closed subgroup H_1 of G. Let H_1 act on M by $\alpha_t(x) = \lambda_t x \lambda_t^*$ for $x \in M$ and $t \in H_1$. Now for M_o we take the linear span of $\{m_f \rho_t | f \in K(G), t \in H\}$. Then again the theorem can be applied and gives

$$\{m_f, \rho_t | f \in K(G), t \in H\}'' \cap \{\lambda_t | t \in H_1\}' = \{m_f, \rho_t | f \in K(H_1 \backslash G),\ t \in H\}''$$

where $K(H_1 \backslash G)$ is the set of continuous complex-valued functions with compact support on the right cosets $H_1 \backslash G$.

Combining the two results we obtain

$$\{m_f, \lambda_t | f \in K(G/H), t \in H_1\}' = \{m_f, \rho_t | f \in K(H_1 \backslash G), t \in H\}''$$

which is a famous result of Takesaki, see also [3,4,5].

REFERENCES

[1] T. DIGERNES, "Poids dual sur un produit croisé" C.R. Acad. Sc. Paris 278A -1974) 937-940.

[2] T. DIGERNES, "Duality for weights on covariant systems and applications" (1975) UCLA Thesis.

[3] O. NIELSEN, "The Mackey-Blattner theorem and Takesaki's generalized commutation relation for locally compact groups". Duke Math. J. 40 (1973) 105-114.

[4] M. RIEFFEL, "Commutation theorems and generalized commutation relations". Bull. Soc. Math. France 104 (1976) 205-224.

[5] M. TAKESAKI, "A generalized commutation relation for the regular representation". Bull. Soc. Math. France 97 (1969) 289-297.

[6] M. TAKESAKI, "Duality for crossed products and the structure of von Neumann algebras of type III" Acta Mathematica 131 (1973) 249-310.

[7] A. VAN DAELE, "Crossed products of von Neumann algebras" Lecture notes (June 1975).

Algebraic Features of Equilibrium States

Daniel Kastler

History (i.e., Maxwell, Gibbs and Boltzmann, as transcribed into quantum mechanics) provides us with the following prescription to describe equilibrium states[1)] $\varphi_{\beta,\mu}$ to the temperature β^{-1} and the chemical potential μ:

$$(1) \qquad \varphi_{\beta,\mu}(a) = \mathrm{Tr}\{e^{-\beta(H-\mu N)}a\}/\mathrm{Tr}\{e^{-\beta(H-\mu N)}\}.$$

Here H is the hamiltonian and N the particle number operator of the system. This procedure (called "Gibbs' Ansatz") is satisfactory for (arbitrarily approximate) numerical results, but inadequate for foundational purposes: (1) requires the system to be "enclosed in a box" (with perfectly reflecting walls, or periodic boundary conditions etc.), whereafter one performs the "thermodynamic limit" (infinite box). Before this limit, the model is highly unphysical (excited states constant in time instead of "return to equilibrium", destruction of invariance under spatial translations, etc.). Besides, one wishes to develop the notions of temperature and chemical potential from first principles.

In order to do this, and treat directly the infinite system, we need a substitute for (1) relevant to the latter situation: this substitute was found by Haag, Hugenholtz and Winnink [1] to be the Kubo-Martin-Schwinger (KMS) condition[2)] formulated in the frame of "C*-systems" $\{\mathfrak{G},R,\tau\}$ (Definition 2 below).

---

1) possibly more than one after the thermodynamic limit, if the latter depends upon boundary conditions (case of phase transitions).

2) proposed by these authors [2] [3] as a boundary condition for the calculation of "Green's functions".

Definition 1. A C*-system $\{\mathfrak{A},G,\tau\}$ is the triple of a C*-algebra $\mathfrak{A}$, a locally compact group G and a representation τ of G into the automorphisms of $\mathfrak{A}$ such that $g \in G \to \tau_g(a)$ is continuous for all $a \in \mathfrak{A}$. Given a τ-invariant state φ of $\mathfrak{A}$, the GNS-construction $\{\pi_\varphi, U_\varphi, \mathcal{H}_\varphi, \Omega_\varphi\}$ of φ is the triple of a *-representation π_φ of $\mathfrak{A}$, a continuous unitary representation U_φ of G, both on $\mathcal{H}_\varphi$, and a vector $\Omega_\varphi \in \mathcal{H}_\varphi$ cyclic for π_φ such that $\varphi(a) = (\Omega_\varphi | \pi_\varphi(a) | \Omega_\varphi)$, $\pi_\varphi(\tau_g(a)) = U_\varphi(g),\ \pi_\varphi(a)U_\varphi(g)$ and $U_\varphi(g)\Omega_\varphi = \Omega_\varphi$, $a \in \mathfrak{A}$, $g \in G$.

Definition 2. A state φ of the C*-system $\{\mathfrak{A},R,\tau\}$ (R the additive reals) is called β-KMS for τ, $\beta \in R$, whenever, to all $a,b \in \mathfrak{A}$, there is a function $z \in \mathbf{C} \to u_{ab}(z) \in C$ holomorphic in the open strip $0 \leq \mathrm{Jm}\, z \leq \beta$, bounded continuous on its boundary, such that

$$\left\{ \begin{array}{l} F_{ab}(t) = \varphi(b\tau_t(a)) = u_{ab}(t) \\ \\ G_{ab}(t) = \varphi(\tau_t(a)b) = u_{ab}(t+i\beta) \end{array} \right. , \quad t \in R. \tag{2}$$

Remark 1. This definition entails that φ is τ-invariant. For τ-invariant states φ, it is equivalent to the following relation between the Fourier transforms of F_{ab}, G_{ab}:

$$\hat{F}_{ab}(E) = e^{\beta E}\, \hat{G}_{ab}(E), \qquad a,b \in \mathfrak{A}, \quad E \in R. \tag{3}$$

The KMS condition of mathematicians corresponds to $\beta = -1$ in Definition 2 [4].

Relation (2) (i.e., essentially $\omega(b\tau_{i\beta}(a)) = \omega(ab)$ for $a,b \in \mathfrak{A}$, a τ-analytic) is easily shown to follow from (1) with

$$\tau_t(a) = e^{i(H-\mu N)t}\, a\, e^{i(H-\mu N)t}, \qquad t \in R,\ a \in \mathfrak{A} \tag{4}$$

i.e.

(5) $$\tau_t = \alpha_t \gamma_{\mu t}$$

where $t \in R \to \alpha_t$ and $g \in T' \to \gamma_g$ are the respective groups of time translations (generated by H) and gauge transformations (generated by -N)[3]:

(6) $$\alpha_t(a) = e^{iHt} a \, e^{-iHt}, \quad \gamma_g(a) = e^{-iNg} a \, e^{iNg}, \qquad t \in R, \; g \in T', \; a \in \mathfrak{F}.$$

Since (2) with τ given by (5) persists through the thermodynamic limit we can replace the complex Gibbs Ansatz + thermodynamic limit by the following requirement [1]:

Let $\{\mathfrak{F}, R \times T^1, \alpha \times \gamma\}$ be the C*-system obtained from the algebra $\mathfrak{F}$ of local fields (the field algebra) acted upon by the direct product of time and gauge. The equilibrium states to the temperature β^{-1} and the chemical potential μ are the states of $\mathfrak{F}$ possessing the β-KMS property for the group $t \to \alpha_t \gamma_{\mu t}$.

This is now the situation to be explained from first principles.

Remark 2. The gauge groups considered here are gauge groups of the first kind, the simplest of which is T^1 as considered above (one species of particles). The general case of a compact (non commutative) gauge group G is of interest in view of groups like SU_3, SU_4, etc. Then $t \to \alpha_t \gamma_{\mu t}$ has to be replaced by $t \to \alpha_t \gamma_{\xi t}$, with $t \to \xi_t$ a continuous one-parameter subgroup of G.

Remark 3. The gauge invariant part $\mathfrak{A}$ of the field algebra[4])

(7) $$\mathfrak{A} = \mathfrak{F}^G = \{A \in \mathfrak{F};\ \gamma_g(A) = A \quad \text{for all} \quad g \in G\}$$

---

3) $\alpha_t \gamma_g = \gamma_g \alpha_t$, $t \in R$, $g \in G$ since H and N commute.

4) We shall denote observables by capitals and fields by low case letters.

is called the algebra of local observables (non-gauge invariant fields are in principle unobservable). In the C*-approach to field theory, $\mathfrak{A}$ is considered as the basic object, the rest of $\mathfrak{F}$ being analytical apparatus constructible from $\mathfrak{A}$ [4]. Since G acts trivially on $\mathfrak{A}$, (5) reduces there to α_t. Denoting by ω the restriction of φ to $\mathfrak{A}$ the above characterization of equilibrium states then splits into

(1) The equilibrium states ω of $\mathfrak{A}$ to temperature β^{-1} are the β-KMS states for the time translations $t \to \alpha_t$ [5].

(2) The extensions φ of such states ω to $\mathfrak{F}$ are β-KMS for a one-parameter mixture $t \to \alpha_t \gamma_{\xi_t}$ of time and gauge.

These statements (1) and (2) correspond respectively to the notions of temperature and chemical potential as treated in I and II below.

I. Temperature (as obtained from stability).

In that paragraph we consider the C*-system $\{\mathfrak{A},R,\alpha\}$ defined by the observable algebra with its dynamics (time translations).

Definition 3. Let $h = h^* \in \mathfrak{A}$. The local perturbation $t \to \alpha_t^{(h)}$ by h of the dynamics $t \to \alpha_t$ is defined by

(8) $$\alpha_t^{(h)}(A) = P_t^{(h)}\alpha_t(A)P_t^{(h)*}, \qquad t \in R, \quad A \in \mathfrak{A},$$

with $P_t^{(h)} \in \mathfrak{A}$ the unitary cocycle determined by

(9) $$\frac{d}{dt} P_t^{(h)} = i\alpha_t(h)P_t^{(h)}, \qquad P_0^{(h)} = I.$$

Remark 4. (9) entails the unitary cocycle property: $P_{t+s}^{(h)} = P_t^{(h)}\alpha_s(P_s^{(h)})$, $P_t^{(h)*} = P_t^{(h)-1} = \alpha_t(P_t^{(h)})$, $s,t \in R$; and also the fact that each α-differentiable $B \in \mathfrak{A}$ is also $\alpha^{(h)}$-differentiable with

---

5) restrictions to $\mathfrak{A}$ of the time translations α_t on $\mathfrak{F}$ (these leave $\mathfrak{A}$ globally invariant since time and gauge commute).

(10) $$\frac{d}{dt}\bigg|_{t=0} \alpha_t^{(h)}(B) = \frac{d}{dt}\bigg|_{t=0} \alpha_t(B) + i[h,B].$$

Relation (10) shows that Definition 3 amounts to "adding the (local) h to the hamiltonian", whence the name "local perturbation".

<u>Definition 4</u>. An α-invariant state ω of $\{\mathfrak{A},R,\alpha\}$ is called <u>stable for local perturbations of the dynamics</u> whenever there is a map $h \in \mathcal{V} \to \omega^{(h)} \in \mathcal{S}$ from a neighbourhood $\mathcal{V}$ of 0 in the self adjoint part of $\mathfrak{A}$ into the state space $\mathcal{S}$ of $\mathfrak{A}$, such that (i) $\omega^{(h)} \circ \alpha_t^{(h)} = \omega^{(h)}$, $t \in R$ (ii) $\omega^{(\lambda h)}(A) \xrightarrow[\lambda=0]{} \omega(A)$, $A \in \mathfrak{A}$ (iii) $\omega^{(h)}(\alpha_t(A)) \xrightarrow[t=\pm\infty]{} \omega(A)$, $A \in \mathfrak{A}$.

$\omega^{(h)}$ is interpreted physically as the perturbed equilibrium state, invariant under $\alpha^{(h)}$, close to ω for small h, and returning to equilibrium.

<u>Theorem 1</u> (Araki [6]). Each α-invariant state ω of $\{\mathfrak{A},R,\alpha\}$ β-KMS for α is stable for local perturbations of the dynamics. The perturbed state $\omega^{(h)}$, contained in the normal folium of ω, is given by the convergent expansion

(11) $$\omega^{(h)}(A) = \frac{\omega(AW^{(h)})}{\omega(W^{(h)})} \quad \text{with}$$

$$W^{(h)} = I + \sum_{n=1}^{\infty} (-1)^n \int_{0\leq s_1 \cdots \leq s_n<1} \alpha_{is_1}(h) \cdots \alpha_{is_n}(h) ds_1 \cdots ds_n.$$

<u>Remark 5</u>. Robinson [7] has shown that expansion [11] can be rewritten as follows in terms of the truncated expectation values ω_n^T of ω:

(12) $$\omega^{(h)}(A) = \omega(A) + \sum_{n=1}^{\infty} \int_{0\leq s_1\leq \cdots \leq s_n<1} \omega_n^T(A\alpha_{is_1}(h) \cdots \alpha_{is_n}(h)) ds_1 \cdots ds_n.$$

<u>Theorem 2</u> (Haag, Trych-Pohlmeyer, Kastler [8]). Assume the following

asymptotically abelian property: there is a norm-dense, α-invariant * subalgebra $\mathfrak{A}_0$ of $\mathfrak{A}$ such that

$$(13) \qquad \int (\|[A,\alpha_t(B)]\| dt < \infty, \qquad A,B \in \mathfrak{A}_0.$$

Let ω be a state of $\mathfrak{A}$ which is (i) α-invariant (ii) stable for local perturbations of the dynamics (iii) hyperclustering of order 4 i.e., to each set $A_1,\dots,A_p \in \mathfrak{A}_0$, $p \le 4$ there are positive C and k with

$$(14) \qquad \left|\omega_p^T(\alpha_{t_1}(A_1) \cdots \alpha_{t_p}(A_p))\right| \le C\{1 + \sup_{i,j} |t_i - t_j|\}^{-1-\delta}.$$

Then, if ω is not a trace

- either ω is β-KMS for α and some $\beta \in R$
- or ω generates a covariant representation (π_ω, U_ω) with a one-sided spectrum of U_ω (energy spectrum).

Remark 6. The second alternative corresponds to a ground state (limiting case $\beta = \infty$). Hyperclustering of order 6 was used in [8] - reduced to order 4 in [9]. Asking ω to be coexisting (i.e., local stability for the system $\{\mathfrak{A},R,\alpha\}$ tensorized by an analogous system in an analogous state (see [10]), one can reduce condition (iii) above to weak clustering ($\Longleftrightarrow$ extremal α-invariance).

Remark 7. One may consider global rather than local perturbations: e.g., Definition 3 still makes sense, under relativistic locality, for h formally replaced by $h_{(f)} = \int f(x)\alpha_x(h)d^3x$, with α_x the space translations and the function f extending to ∞. One can show either on models [11] or generally via appropriate clustering assumptions (Hugenholtz, Mebkhout and Kastler, work in progress), the validity of stability for global perturbations (for instance for $\alpha^{(h_f)}$ with $f \equiv 1$ at low activity).

Such perturbations also allow to treat media in motion [10].

II. Chemical potential (as obtained by extending from $\mathfrak{A}$ to $\mathfrak{F}$)

We now work with the C*-system $\{\mathfrak{F}, R \times G, \alpha \times \gamma\}$, G compact, $\mathfrak{A} = \mathfrak{F}^G$, and investigate extensions to $\mathfrak{F}$ of extremal β-KMS states ω of $\mathfrak{A}$ for α. We know three methods for treating the extension problem: two presented in [12] the second of which based on [15]), one in [13].

Theorem 3 (Araki, Haag, Takesaki and Kastler [11]). Let $\{\mathfrak{F}, R \times G, \alpha \times \gamma\}$ be a C*-system with $R \times G$ the direct product of the time axis and a compact gauge group G and assume $\{\mathfrak{F}, R, \alpha\}$ asymptotically abelian

$$\|[a,\alpha_t(b)]\| \xrightarrow[t=\infty]{} 0, \quad a,b \in \mathfrak{F}, \tag{15}$$

(for the adaptation to Fermi fields see [12]). Then

(i) each extremal α-invariant state ω of $\mathfrak{A}$ has an extremal α-invariant extension φ to $\mathfrak{F}$. Furthermore two such extensions φ_1, φ_2 are such that $\varphi_2 = \varphi_1 \circ \gamma_g$ for some $g \in G$.

(ii) let ω be a state of $\mathfrak{A}$ extremal β-KMS for α and assume ω to be faithful $(\varphi(A^*A) = 0,\ A \in \mathfrak{A} \Rightarrow A = 0)$. If φ is an extremal α-invariant extension of ω to $\mathfrak{F}$, φ is β-KMS for a one-parameter group of the type $t \to \alpha_t \gamma_{\xi_t}$, $t \to \xi_t$ a continuous one-parameter subgroup of the center of the stabilizer G_φ of φ $(G_\varphi = \{g \in G;\ \varphi \circ \gamma_g = \varphi\})$

(iii) in addition to the assumptions of (ii) take $G = T^1$ (so that $\xi_t = \mu t$ for some $\mu \in R$) and assume $\mathfrak{A}$ to possess a relativistic local structure. The chemical potential μ is characterizable as follows: with ρ a localized automorphism of $\mathfrak{A}$ and $v_t \in \mathfrak{A}$ the

---

6) The automorphism ρ of $\mathfrak{A}$ is localized (in the region R) (see [5]) if it reduces to the identity on the algebra corresponding to the points space like to all points of R.

unitary cocycle relating ρ to its time conjugates:

(16) $$\rho^{-1} \circ \alpha_t \circ \rho = \mathrm{Ad} v_t \circ \alpha_t, \qquad t \in R,$$

one has that (a) $\omega \circ \rho$ is quasi-equivalent to ω (b) the cocycle Radon-Nikodym derivative of $\omega \circ \rho$ w.r.t. ω (see [14]) is related with v_t as follows:

(17) $$(D(\omega \circ \rho):D\omega)_t = e^{in\beta(\mu+c)t}\, \pi_\omega(v_{-\beta t}),$$

where c is a real constant independent of ω and π_ω is the representation generated by ω.

Theorem 4 (Araki, Kishimoto [13]).[7] Let $\{\mathfrak{F}, R \times G, \alpha \times \gamma\}$ be a C*-system with G compact abelian. Set $\varepsilon_\gamma = \int \overline{\gamma(g)}dg$, $\gamma \in \hat{G}$, and assume the existence of a generating set Δ of $\hat{G}$ such that:

(A) For each $\gamma \in \Delta$ there is a norm-bounded sequence $\{b_n\} \subset \varepsilon_\gamma(\mathfrak{F})$ with the properties: the $\alpha_t(b_n)$ are equicontinuous in t; $b_n^* b_n + b_n b_n^* \geq 1$ for all n; the commutators $[b_n, A]$, $[b_n^*, \alpha_t(b_n), A]$, $[\alpha_t(b_n) b_n^*, A]$ all tend to 0 in norm $n \to \infty$.

Moreover, for a state ω of $\mathfrak{A} = \mathfrak{F}^G$ we consider the following assumption

(B) For each $\gamma \in \Delta$, $\sup_n s_c\{\pi_\omega(b_n^* b_n)\} = \sup_n s_c\{\pi_\omega(b_n b_n^*)\} = 1$, where $\{b_n\}$ is the sequence considered in (A) and s_c denotes the central support in $\pi_\omega(\mathfrak{A})''$.

Then, with Σ the set of one-parameter subgroups of $R \times G$ of the form $t \to \alpha_t \gamma_{\xi t}$, $t \to \xi_t$ a continuous one-parameter subgroup of G, denoting by

- $\mathcal{S}_\beta(\mathfrak{A})$ the set of states of $\mathfrak{A}$ satisfying (B) and β-KMS for $t \to \alpha_t$,
- Ext $\mathcal{S}_\beta(\mathfrak{A})$ ———————————— extremal β-KMS for $t \to \alpha_t$,
- $\mathcal{S}_\beta(\mathfrak{F})$ the set of β-KMS states of $\mathfrak{F}$ for a one-parameter group in Σ,

---

7) We modified the notations of [13] for uniformity. Our $\mathfrak{F}$, resp $\mathfrak{A}$, α, γ, φ, ω is their $\mathfrak{A}$, resp $\mathfrak{A}^G$, ρ, α, $\hat{\varphi}$ and φ.

- $\mathcal{S}_\beta^G(\mathfrak{F})$ the subset of γ-invariant states in $\mathcal{S}_\beta(\mathfrak{F})$,
- Ext $\mathcal{S}_\beta^G(\mathfrak{F})$ the set of extremal elements of $\mathcal{S}_\beta^G(\mathfrak{F})$,

we have that

(i) $\omega \in \text{Ext}\ \mathcal{S}_\beta(\mathfrak{A})$ implies $\overline{\omega} = \omega \circ \varepsilon_1 \in \text{Ext}\ \mathcal{S}_\beta(\mathfrak{F})$.

(ii) $\varphi \in \mathcal{S}_\beta(\mathfrak{F})$ implies $\varphi|_{\mathfrak{A}} \in \mathcal{S}_\beta(\mathfrak{A})$. Further if φ is primary, $\varphi|_{\mathfrak{A}} \in \text{Ext}\ \mathcal{S}_\beta(\mathfrak{A})$; if $\varphi \in \mathcal{S}_\beta^G(\mathfrak{F})$, $\varphi = \varphi|_{\mathfrak{A}} \circ \varepsilon_1$; and if $\varphi \in \text{Ext}\ \mathcal{S}_\beta^G(\mathfrak{F})$, $\varphi|_{\mathfrak{A}} \in \text{Ext}\ \mathcal{S}_\beta(\mathfrak{A})$.

(iii) one has a bijection $\omega \in \text{Ext}\ \mathcal{S}_\beta(\mathfrak{A}) \leftrightarrow \varphi \in \text{Ext}\ \mathcal{S}_\beta^G(\mathfrak{F})$ by taking $\omega = \varphi|_{\mathfrak{A}}$ and $\varphi = \overline{\omega} = \omega \circ \varepsilon_1$.

(iv) for any $\omega \in \text{Ext}\ \mathcal{S}_\beta(\mathfrak{A})$ there is an extremal β-KMS state for $t \to \rho_t$ for some $\rho \in \Sigma$ with $\varphi|_{\mathfrak{A}} = \omega$. If $\varphi \circ \gamma_{g_1} \neq \varphi\gamma_{g_2}$, $g_1, g_2 \in G$, $\varphi \circ \gamma_{g_1}$ and $\varphi \circ \gamma_{g_2}$ are disjoint. The central decomposition of $\overline{\omega} = \omega \circ \varepsilon_1$ is given by

$$\omega = \int \varphi \circ \gamma_g dg.$$

Remark 7. The faithfulness assumption in Theorem 3(ii) (automatic if $\mathfrak{A}$ is simple) and assumption (B) in Theorem 4, are for ruling out the occurrence of "vacuum like properties" in gauge, see [12] and [13].

Remark 8. In [2] we treat the more general case of an asymptotically abelian group τ commuting with α (e.g., space translations) and an extremal τ-invariant state of $\mathfrak{A}$. We give here the version of [12] which seems adapted to relativistic fields, since [13] seems preferable for non relativistic models (e.g., spin systems).

Remark 9. For an investigation of the representation π_ω of $\mathfrak{F}$ generated by the "average" $\overline{\omega}$ we refer to [13], Theorem 2, and Section III of [12].

References

[1] R. Haag, N. Hugenholtz and M. Winnink, On the Equilibrium States in Quantum Statistical Mechanics, Comm. Math. Phys, 5, 215 (1967).

[2] R. Kubo, Statistical-mechanical Theory of Irreversible Processes. I. General Theory and simple Applications to magnetic and Conduction Problems, J. Phys. Soc. Japan, 12, 570 (1957).

[3] P. C. Martin and J. Schwinger, Theory of Many Particle Systems, Phys. Rev., 115, 1342 (1959).

[4] M. Takesaki, Tomita's Theory of Modular Hilbert Algebras and its Applications. Springer Lecture Notes in Math., No. 128.

[5] S. Doplicher, R. Haag, J. Roberts, Fields, Observables and gauge Transformations I and II, Comm. Math. Phys., 13, 1 (1969); 15 173 (1969).

[6] H. Araki, Relative Hamiltonian for faithful normal States of a Von Neumann algebra. Pub. Res. Inst. Math. Sci. Kyoto University 9, No. 1, 165 (1973).

[7] Derek W. Robinson, Perturbations Expansions of KMS States, CPT-CNRS Preprint 74/P. 633, Marseille (1974).

[8] R. Haag, D. Kastler and E. Trych-Pohlmeyer, Stability and Equilibrium States, Comm. Math. Phys., 38, 173 (1974).

[9] O. Bratteli, D. Kastler, Relaxing the clustering condition in the Derivation of the KMS Property, Comm. Math. Phys.

[10] D. Kastler, Foundations of Equilibrium Statistical Mechanics, U.C.L.A. Lecture Notes, (April 1977).

[11] R. Haag, E. Trych-Pohlmeyer. Private Communication.

[12] H. Araki, R. Haag, D. Kastler, M. Takesaki. Extension of States and Chemical Potential. Comm. Math. Phys, 53, 97 (1977).

[13] H. Araki, A. Kiskimoto. Symmetry and Equilibrium States, Comm. Math. Phys. 52, 211 (1977).

[14] A. Connes. Une classification des facteurs de type III. Ann. Scient. Ecole Norm. Sup., 6, 133 (1973).

[15] J. Roberts. Cross products of von Neumann algebras by group duals. Proceedings of the Conference on C*-algebras - Symposia Mathematica XX 333 (1976).

MINIMAL DILATIONS OF CP-FLOWS(*)

Gerard G. Emch

Dpts. of Mathematics and of Physics, University of Rochester.

We consider the following problem. Given $\{\mathcal{N}, \phi, \gamma\}$ where $\mathcal{N}$ is a von Neumann algebra; ϕ is a faithful normal state on $\mathcal{N}$; $\{\gamma_t \mid t \in R^+\}$ is a semigroup of completely positive maps [18] of $\mathcal{N}$ onto itself such that $\phi \cdot \gamma_t = \phi$, $\gamma_t[I] = I$ and $\lim_{t\to\infty} \psi \cdot \gamma_t = \phi$ for all ψ in $\mathcal{N}_*^{+,1}$. Find $\{\underline{\mathcal{N}}, \underline{\phi}, \underline{\alpha}, i, \mathcal{E}\}$ where $\underline{\mathcal{N}}$ is a von Neumann algebra; $\underline{\phi}$ is a faithful normal state on $\underline{\mathcal{N}}$; $\{\underline{\alpha}_t \mid t \in R\}$ is a group of automorphisms of $\underline{\mathcal{N}}$ such that $\underline{\phi} \cdot \underline{\alpha}_t = \underline{\phi}$; i is an injective *-homomorphism of $\mathcal{N}$ into $\underline{\mathcal{N}}$; $\mathcal{E}$ is a faithful normal conditional expectation from $\underline{\mathcal{N}}$ onto $\mathcal{N}$ such that $\phi \cdot \mathcal{E} = \underline{\phi}$, $\mathcal{E} \circ \underline{\alpha}_t \cdot i = \gamma_t$ for all t in R^+, and $\{\underline{\alpha}_t \circ i[\mathcal{N}] \mid t \in R\}'' = \underline{\mathcal{N}}$. When such a minimal conservative dilation of the CP-flow $\{\mathcal{N}, \phi, \gamma\}$ exists, we further ask whether the condition $\lim_{t\to\infty} \psi \cdot \gamma_t = \phi$ implies any ergodic properties on $\{\underline{\mathcal{N}}, \underline{\phi}, \underline{\alpha}\}$ such as mixing, asymptotic abelianness, Lebesgue spectrum or strict positivity of a properly defined dynamical entropy.

The motivation for this problem stems from the following physical interpretation. $\mathcal{N}$ is an algebra of observables on a dissipative physical system whose approach to equilibrium ϕ is governed by the dynamical semigroup γ. The assumptions that $\gamma_t[I] = I$ and γ_t positive are made to ensure that whenever ψ is a state on $\mathcal{N}$ so is $\psi \cdot \gamma_t$. Complete positivity appears then as a consistency requirement on composite systems [1,5,13,14,15,16]. The semigroup property corresponds to situations described by memoryless transport equations, such as the Bloch equation [15] for a spin system, or the diffusion equation for a particle in a harmonic well [4,9]. We thus ask whether the dissipative system $\{\mathcal{N}, \phi, \gamma\}$ can be considered as a subsystem (or partial description) of an underlying conservative system in canonical equilibrium $\underline{\phi}$; $\underline{\mathcal{N}}$ describes the observable-algebra of this larger system where the evolution is governed by the dynamical group $\underline{\alpha}$. The last condition, namely $\{\underline{\alpha}_t \circ i[\mathcal{N}] \mid t \in R\}'' = \underline{\mathcal{N}}$ can always be imposed a posteriori once the other conditions are satisfied; it is thus a necessary condition for the dilation to be canonical. Finally,

(*) Research supported in part by the NSF (Grant MCS 76-07286)

: should be emphasized that we only require $\phi \cdot \underline{\alpha}_t = \underline{\Phi}$, and not that $\underline{\alpha}$ be the modular .9] group $\underline{\sigma}$ associated to $\underline{\phi}$. The reason is that we have in mind [12] weak-coupling[5] ıeories where $\underline{\alpha}$ describes (in the interaction picture) the long-time cumulative effect the interaction responsible for the approach to the equilibrium state $\underline{\phi}$ which is ıS for the free evolution $\underline{\sigma}$.

: should observe that the restrictions imposed by these physical considerations are ıite demanding. This is particularly true for the existence of the distinguished state . Indeed the existence of a dilation $\{\underline{\mathcal{N}}, \underline{\alpha}\}$ of $\{\mathcal{N}, \gamma\}$ has recently been established, ıd this even if one weakens the assumption $\gamma_s \gamma_t = \gamma_{s+t}$ to the assumption that γ is ıly indexed [13,6] by elements of a locally compact group G (here R).

ı the setting of classical theories where $\mathcal{N}$ is realized as $\mathcal{L}^\infty(\Omega, \Sigma, \mu)$ acting on $=\mathcal{L}^2(\Omega, \Sigma, \mu)$ with (Ω, Σ, μ) a probability space, γ induces a <u>Markov semigroup</u> on $\mathcal{H}$: $| t \in R^+\}$ with $P_t N \Phi = \gamma_t[N]\Phi$ where N runs over $\mathcal{N}$ and Φ is $\Phi : \omega \in \Omega \mapsto 1 \in \mathbb{C}$.

ı this case, our problem reduces to the classical <u>Kolmogorov-Daniell</u> [20] reconstruction a <u>Markov process</u> indexed by R, with state space (Ω, Σ) and underlying probability ıace $(\underline{\Omega}, \underline{\Sigma}, \underline{\mu})$. We have then $\underline{\mathcal{N}} = \mathcal{L}^\infty(\underline{\Omega}, \underline{\Sigma}, \underline{\mu})$; $\langle \underline{\phi} ; \underline{N} \rangle = \underline{\mu}(\underline{N})$; $\underline{\alpha}_t$ induced by the ıift on $\underline{\Omega} = \Omega^R$; i : $N \in \mathcal{N} \mapsto N \cdot \pi_c \in \underline{\mathcal{N}}$ where $\pi_c(\underline{\omega}) = \underline{\omega}(0)$; and $\mathcal{E}$: $\underline{N} \in \underline{\mathcal{N}} \mapsto E(\underline{N} | \sigma(X_0)) \in \mathcal{N}$

ı some extend this construction can be transposed to the non-commutative setting as ıllows. For every finite ordered subset $\Lambda = \{t_1, \ldots, t_n\} \subset R^+$, define on the n-fold ırtesian product $\mathcal{N} \otimes \ldots \otimes \mathcal{N}$ the sesquilinear form :

$$\langle \sum_j N_{1,j} \otimes \ldots \otimes N_{n,j}, \sum_k M_{1,k} \otimes \ldots \otimes M_{n,k} \rangle_\Lambda =$$

$$\sum_{j,k} \langle \Phi ; M^*_{n,k} \gamma_{t_n - t_{n-1}}[M^*_{n-1,k} \ldots \gamma_{t_2 - t_1}[M^*_{1,k} N_{1,j}] \ldots N_{n-1,j}] N_{n,j} \rangle$$

ıe complete positivity of γ_t ensures that $\langle \cdot , \cdot \rangle_\Lambda$ is positive. Let K_Λ be its kernel, ıd $\mathcal{H}_\Lambda$ be the Hilbert space completion of $\mathcal{N} \otimes \ldots \otimes \mathcal{N} / K_\Lambda$. Denote by Φ_Λ the equivalence ıass of $I \otimes \ldots \otimes I$. The conditions $\gamma_t[I] = I$ and $\gamma_s \gamma_t = \gamma_{s+t}$ ensure consistency ı the sense that for any pair Λ_1, Λ_2 of finite ordered sets in R^+, with $\Lambda_1 \subset \Lambda_2$ ıere exists an isometry from $\mathcal{H}_{\Lambda_1}$ into $\mathcal{H}_{\Lambda_2}$ such that $V_{21} \Phi_{\Lambda_1} = \Phi_{\Lambda_2}$. Hence

$\{\mathcal{H}_\Lambda, \Phi_\Lambda | \Lambda \in \mathcal{F}\}$ is a injective family indexed by $\mathcal{F}$, the set of all finite subsets of R^+ directed by inclusion. Let then $(\underline{\mathcal{H}}, \underline{\Phi})$ be its direct limit, and V_Λ be the resulting isometry of $\mathcal{H}_\Lambda$ into $\underline{\mathcal{H}}$. $\phi \circ \gamma_t = \phi$ then implies that the shift $s \mapsto s+t$ on R^+ induces a one-parameter semi-group of isometries $\underline{U}_t$ of $\underline{\mathcal{H}}$ such that $\underline{U}_t \underline{\Phi} = \underline{\Phi}$. For any $\{0\} \subseteq \Lambda \in \mathcal{F}$ define $i_\Lambda : N \in \mathcal{N} \mapsto \mathcal{B}(\mathcal{H}_\Lambda)$ by $N_0 \otimes N_{t_2} \otimes \ldots \mapsto (NN_0) \otimes N_{t_2} \otimes \ldots$. In case $\Lambda = \{0, t\}$ this reduces to the Stinespring representation of $\mathcal{N}$ induced by γ_t. Let now denote by i the extension of $\{i_\Lambda | \Lambda \in \mathcal{F}\}$ to $\underline{\mathcal{H}}$ and by V the isometry $V_{\{0\}}$. One then checks that $\gamma_t[N] = V^* \underline{U}_t^* \; i(N) \underline{U}_t \; V$ for all t in R^+ and all N in $\mathcal{N}$. Compare with [1,14,1?] This result falls short of being a solution to our problem, specifically because $\underline{U}_t$ is a partial isometry rather than a unitary operator. In the classical case the above construction corresponds to that of a Markov process indexed by R^+ ; its extension from R^+ to R (i.e. from Ω^{R^+} to Ω^R) offers no difficulties, and $\underline{U}_t$ becomes a unitary operator from which we can complete the construction. In the general case however, the substitution of R^+ by R at the beginning of the above construction would lead to an injective map i which is a representation, but not a *-representation unless $\mathcal{N}$ is abelian.

In view of this, it is gratifying to know that a complete solution to our problem still exists in some particular cases of physical significance. Let indeed $\{\mathfrak{H}, B, S\}$ be any triple formed by a (separable) Hilbert space $\mathfrak{H}$, a self-adjoint operator $0 < B < I$ acting on $\mathfrak{H}$, and a strongly continuous semigroup of bounded operators $\{S_t | t \in R^+\}$ acting on $\mathfrak{H}$ such that $s\text{-}\lim_{t\to\infty} S_t$ exists, and $\|h\|^2 - \|S_t h\|^2 \geq \|h\|_B^2 - \|S_t h\|_B^2 > 0$ for every $t > 0$ and every h in $\mathfrak{H}$ (where we wrote $\|h\|_B^2$ for (Bh,h)). Such a triple will be refered to as a <u>completely contractive semigroup</u>. Let now $\phi : h \in \mathfrak{H} \mapsto \exp\{-\|h\|^2/4\}$, and V be the corresponding representation of the CCR with $V(h_1)V(h_2) = V(h_1+h_2)\exp\{i\,\mathrm{Im}(Bh_1,h_2)/2\}$. Then $\gamma_t[V(h)] = V(S_t h)\,\phi(h)/\phi(S_t h)$ extends to a semigroup of completely positive maps (compare with [7]) on $\mathcal{N} = \{V(h) \mid h \in \mathfrak{H}\}''$, with $\gamma_t[I] = I$, $\phi \circ \gamma_t = \phi$ and $\lim_{t\to\infty} \psi \cdot \gamma_t = \phi$ for all ψ in $\mathcal{N}_*$. The properties of the Nagy extension of $(\mathfrak{H}, S)$ lead to the following :

<u>Theorem</u> [11] : The <u>CP-flow</u> $\{\mathcal{N}, \phi, \gamma\}$ constructed from any completely contractive semigroup $\{\mathfrak{H}, B, S\}$ <u>admits a minimal conservative dilation</u> $\{\mathcal{N}, \psi, \alpha, i, \mathcal{E}\}$. Let $\mathcal{A} = \{\alpha_t \cdot i[\mathcal{N}] | t \leq 0\}''$. Then : (i) $\mathcal{A} \subseteq \alpha_t[\mathcal{A}]$ for all $t \geq 0$; (ii) $\bigcap_t \alpha_t[\mathcal{A}] = \mathbb{C}\, I$, and

$_t\,\alpha_t[\mathcal{A}] = \mathcal{N}$;(iii) $\mathcal{A}$ is stable with respect to the modular group σ associated to Φ ; e. $\{\mathcal{N}, \Phi, \alpha, \mathcal{A}\}$ is a generalized K-flow [10]. (iv) the von Neumann algebra of fixed ints of $\mathcal{N}$ for $\{\alpha_t | t\in R\}$ reduces to $\mathbb{C}$ I , i.e. $\{\mathcal{N}, \Phi, \alpha\}$ is ergodic. (v) α is implemented $\{U_t \mid t\in R\}$ with Lebesgue spectrum, i.e. there exists a decomposition $\mathcal{H} = \oplus_{n=0}^{\infty} \mathcal{H}^{(n)}$ ducing U_t , with $\mathcal{H}^{(o)}$ one-dimensional, $U_t^{(o)} = I$, and $\mathcal{H}^{(n)} \cong \mathcal{L}^2(R,dx)$, $U_t^{(n)}$ unit-ily equivalent to $V_t\Psi(x) = \Psi(x-t)$. (vi) $\mathcal{N}$ is a factor of type III_λ , with all lues $0<\lambda\leq 1$ accessible by a proper choice of ($\mathfrak{H}$,B,S). (vii)The generalized dynamical tropy [8,10] of $\{\mathcal{N}, \Phi, \alpha\}$ is strictly positive, unless the centralizer $\mathcal{N}_o$ of $\{\mathcal{N}, \Phi\}$ duces to C I . Moreover, let $\mathcal{O}\!\mathcal{C}$ (resp. $\mathcal{C}$) be the C*-algebra generated by $\{\alpha_t[\mathcal{A}] | t\in R\}$ esp. $\{\alpha_t[\mathcal{N}\cap\mathcal{A}'] | t\in R\}$). Then: (viii) $\mathcal{O}\!\mathcal{C}$ and $\mathcal{C}$ are strongly dense in $\mathcal{N}$, and for every A $\mathcal{O}\!\mathcal{C}$ and C in $\mathcal{C}$, $\|[A, \alpha_t[C]]\|$ tends to zero as t approaches infinity, i.e. $\{\mathcal{N}, \alpha\}$ is norm-ymptotically abelian in the sense of Araki [2]. (ix) $w\text{-}\lim_{t\to\infty} \alpha_t[A] = \langle\Phi;A\rangle I$ for ery A in $\mathcal{O}\!\mathcal{C}$, and $w\text{-}\lim_{t\to-\infty} \alpha_t[C] = \langle\Phi;C\rangle I$ for every C in $\mathcal{C}$. The dilation is nally Markovian in the sense that : (x) for any $\{X_k , Y_k \; k=1,2,\ldots,n\}\subset\mathcal{N}$ and any nite ordered set $\{t_1,t_2,\ldots,t_n\}$ in R^+: $\gamma_{t_1}[X_1\,\gamma_{t_2}[X_2 \ldots \gamma_{t_n}[X_nY_n] \ldots Y_2]Y_1] =$ $(\alpha_{t_1}[i(X_1)\,[\alpha_{t_2}[i(X_2) \ldots \alpha_{t_n}[i(X_nY_n)] \ldots i(Y_2)]\,i(Y_1)])$.

BLIOGRAPHY

L.Accardi, Adv.Math 20 (1976) 329-366
H.Araki,Commun.math.Phys.28 (1972) 267-277
A.Connes & E.Størmer, Acta Math. 134 (1975) 289-306
E.B.Davies,Commun.math.Phys. 27 (1972) 309-325
E.B.Davies,Quantum Theory of Open Systems, Academic Press, 1976
E.B.Davies,preprint, Oxford, 1976
B.Demoen,P.Vanheuverzwijn & A.Verbeure,preprint,Leuwen,1976
G.G.Emch,Z.Wahrscheinlichkeitstheorie verw.Gebiete 29 (1974) 241-252
G.G.Emch,Acta Phys.Austriaca,Suppl.XV (1976) 79-131; and in Physical Reality & Mathematical Description, C.P.Enz & J.Mehra,eds.,Reidel, 1974,pp.477-493
G.G.Emch,Commun.math.Phys. 49 (1976) 191-215 ; Journ.Funct.Analysis 19 (1975) 1-12
G.G.Emch,S.Albeverio & J.P.Eckmann,preprint,Genève,1977
G.G.Emch,preprint, Rochester, 1977
D.E.Evans,Commun.math.Phys. 48 (1976) 15-22
D.E.Evans & J.T.Lewis,Commun.math.Phys. 50 (1976) 219-227
V.Gorini,A.Frigerio,M.Verri,A.Kossakowski & E.C.G.Sudarshan,preprint,Texas, 1976
G.Lindblad,Commun.math.Phys. 48 (1976) 119-130
G.Lindblad,preprint,Stockholm,1977; and Letters Math.Phys. 1 (1976) 219-224
W.F.Stinespring,Proc.Amer.Math.Soc. 6 (1955) 211-216
M.Takesaki, Lecture Notes in Mathematics #128 , Springer, 1970
H.G.Tucker, A Graduate Course in Probability, Academic Press, 1967

Resistance Inequalities for the Isotropic Heisenberg Model

R. T. Powers
University of Pennsylvania
Philadelphia, PA 19174

The purpose of this talk is to point out the relation between the isotropic Heisenberg model and electical resistance. It is our contention that the problem of whether certain systems have a first order phase transition can be determined from the resistance properties of the system. We begin with a discussion of the Heisenberg model, followed by a discussion of the resistance and end with showing the connection between the two.

In this talk we will confine our attention to spin $\frac{1}{2}$ although all the results stated have generalizations to higher spin. To describe the spin at a single spin $\frac{1}{2}$ particle one specifies a state ω of a (2×2)-matrix algebra N. The algebra N is spanned as a linear space by the identity I and the three Pauli spin matrices.

$$I = \begin{pmatrix} 1 & 0 \\ 0 & 1 \end{pmatrix} \qquad \sigma_x = \begin{pmatrix} 0 & 1 \\ 1 & 0 \end{pmatrix}$$
$$\sigma_y = \begin{pmatrix} 0 & -i \\ i & 0 \end{pmatrix} \qquad \sigma_z = \begin{pmatrix} 1 & 0 \\ 0 & -1 \end{pmatrix} \tag{1}$$

(We denote the triple $(\sigma_x,\sigma_y,\sigma_z)$ by $\vec{\sigma}$. A state ω of N is determined by the three numbers $(\omega(\sigma_x),\omega(\sigma_y),\omega(\sigma_z)) = (a_x,a_y,a_z) = \vec{a}$. From the fact that ω is a state and therefore positive we have $|\vec{a}| \leq 1$.

To describe the spin of two particles of spin $\frac{1}{2}$ one specifies a state ω of $N_1 \otimes N_2$ where N_1 and N_2 are (2×2)-matrix algebras. N_1 is the span of the identity $I = I \otimes I$ and the Pauli spin matrices $\vec{\sigma}_j = \vec{\sigma} \otimes I$ and N_2 is spanned by the matrices $\vec{\sigma}_2 = I \otimes \vec{\sigma}$. To describe the spin of a countably infinite number of span $\frac{1}{2}$ particles one specifies a state ω of $\mathfrak{A} = \otimes_{i \in L} N_i$ where $\mathcal{L}$ is a set labelling the particle and N_i is a (2×2)-matrix algebra. $\mathfrak{A}$ is the infinite tensor product of (2×2)-matrix algebras and is called a UHF algebra.

We consider a particular dynamics of this C^*-algebra $\mathfrak{A}$. A dynamics is a strongly continuous one parameter group of $*$-automorphisms of $\mathfrak{A}$. We are interested in the dynamic associated with the Heisenberg model. This dynamics is specified by giving a Hamiltonian H of $\mathcal{L}$

$$H = \Sigma_{(i,j) \in G} J(ij)(I - \vec{\sigma}_i \cdot \vec{\sigma}_j) \tag{2}$$

where the sum is taken over all lines (i,j) of a graph G. The graph G is simply a set of pairs of points (or vertices) (i,j) in $\mathcal{L}$. The graphs of greatest interest to physicists are where $\mathcal{L} = \mathbb{Z}^n$ and G consists of all lines connecting nearest neighbors of $\mathcal{L}$. The number $J(ij)$ are positive. The assumption $J(ij) > 0$ corresponds to the statement; the interaction is ferromagnetic. The expression $\vec{\sigma}_i \cdot \vec{\sigma}_j$ is of course short hand for $\sigma_{jx}\sigma_{ix} + \sigma_{iy}\sigma_{jy} \quad \sigma_{iz}\sigma_{jz}$.

We assume the numbers $J(ij)$ satisfy the inequality

$$\Sigma_{j \in G(i)} J(ij) \leq K \ . \tag{3}$$

where K is a constant and $G(i)$ is the set of vertices $j \in \mathcal{L}$ connected in i by a line (ij) of G. The expression (2) for the Hamiltonian is in general not well defined but with assumption (3) H can be used to define a $*$-derivation δ given by

$$\delta(A) = i \sum_{(ij)\in G} J(ij)[I - \vec{\sigma}_i \cdot \vec{\sigma}_j, A] . \tag{4}$$

From (3) it follows that the above sum converges for all $A \in \mathfrak{A}_0$, with $\mathfrak{A}_0$ in the $*$-subalgebra of $\mathfrak{A}$ spanned by all polynomials in the $(\vec{\sigma}_i; \ i \in \mathcal{L})$. From the rapidly developing theory of unbound $*$-derivations discussed in Professor Sakai's lecture it follows that the closure of δ in the generator of a strongly continuous reparameter group of $*$-automorphisms α_t. We will call α_t a $*$-automorphism group associated with the Hamiltonian H.

We will be interested in two types of states associated with the Hamiltonian H. One type is states of finite energy. These are simply states ω of $\mathfrak{A}$ satisfying

$$\sum_{(ij)\in G} J(ij) \ \omega(I - \vec{\sigma}_i \cdot \vec{\sigma}_j) < \infty . \tag{5}$$

The second type is (α_t,β) KMS states of $\mathfrak{A}$. If G is finite and $\mathfrak{A}$ is $(2^n \times 2^n)$-matrix algebra a (α_t,β)-KMS state of A is the state given by

$$\omega(A) = \frac{tr(A \ e^{-\beta H})}{tr(e^{-\beta H})} \tag{6}$$

where tr is the trace of $\mathfrak{A}$. When $\mathcal{L}$ is infinite (α_t,β)-KMS states use a generalization of the above notion to infinite systems. We refer to Professor Araki's lecture for a description of KMS states.

The Heisenberg model was invented to explain, or give a model for, the existence of magnets. In an iron crystal the spin of neighboring iron atoms interact in accordance with Hamiltonian (1). Naturally this is simplification of an actual iron crystal but the model is believed to be realistic enough to explain why magnets exist in nature. The mathematical property corresponding to the existence of a magnet is long range order. What we want to prove is that if ω is on (α_t,β) KMS state then for certain physically reasonable Hamiltonians ω has long range order, i.e. $\omega(\vec{\sigma}_i \cdot \vec{\sigma}_j) \geq \varepsilon > 0$ for all $i,j \in \mathcal{L}$. Long range order corresponds to an alignment of spins and physically the alignment of the spins of 10^{23} atoms results in a magnet.

Quite recently some very important work has been done on the Heisenberg model. Fröhlich, Simon and Spencer have proved the existence of long range order for the 3 dimensional classical Heisenberg model [1]. Their arguments rely heavily on the symmetry of 3 dimensional Heisenberg model. It is my hope that by using ideas connected with resistance their results can be extended to more general systems.

Next we discuss the calculation of the resistance of an electrical network consisting of resistors. Consider a finite graph G with vertices $\mathcal{L}$. We image that for each line $(i,j) \in G$ there is resistor of $J(ij)^{-1}$ ohms. Given two vertices $i_0, j_0 \in \mathcal{L}$ we wish to calculate the electrical resistance $R(i_0,j_0)$ between i_0 and j_0. We imagine injecting one ampere of current at i_0 and extracting one ampere of current at j_0. The current flows from i_0 to j_0 along the lines (ij) of G. The current flow is governed by two laws. One is Ohn's law which states that the voltage difference between two vertices i and j connected by a line $(ij) \in G$ is given by

$$V(i) - V(j) = J(ij)^{-1} I(ij) \tag{7}$$

where $I(ij)$ is the current flowing along the line (ij): (then $I(ij) = -I(ji)$). The second law is the Kirchhoff law which states that the total current flowing into a vertice k is zero unless $k = i_0$ or $k = j_0$. The Kirchhoff law can be stated

$$\Sigma_{j \in G(ij)} I(ij) = \delta_{i_0}(i) - \delta_{j_0}(i) \tag{8}$$

where $\delta_{i_0}(k) = 0$ if $k \neq i_0$ and $\delta_{i_0}(i_0) = 1$.

These two equations can be combined as follows. If f is a function defined on $\mathcal{L}$ we define Δf by

$$(\Delta f)(i) = \Sigma_{j \in G(i)} J(ij)(f(j) - f(i)) \ . \tag{9}$$

We say a function f is harmonic if $\Delta f = 0$. If G is finite one can easily show that the only harmonic functions are constant. To calculate the resistance between $i_0, j_0 \in \mathcal{L}$ one solves the equation

$$-\Delta V = \delta_{i_0} - \delta_{j_0} \ . \tag{10}$$

The current flow $I(ij) = J(ij)(V(i) - V(j))$ for $(i,j) \in G$ satisfies Ohm's law and the Kirchhoff law. The resistance $R(i_0,j_0)$ is given by the voltage difference

$$R(i_0,j_0) = V(i_0) - V(j_0) \ . \tag{11}$$

Since two different solutions to equation (10) differ by a constant function equation (10) determines the resistance $R(i_0,j_0)$ uniquely.

For infinite graphs we define the resistance as the limit

$$R(i_0,j_0) = \lim R_n(i_0,j_0)$$

where $R_n(i_0,j_0)$ is the resistance as calculated from a finite subgraph G_n of G and the graphs G_n increase up to G as $n \to \infty$. One can show that the limit is independent of the increasing sequence G_n.

An equivalent definition of resistance is as follows. Given G is finite or infinite graph with vertices $\mathcal{L}$ and a function f on $\mathcal{L}$ define

$$Q(f) = \Sigma_{(ij)\in G} J(ij)(f(i) - f(j))^2 . \tag{12}$$

Then the resistance $R(i,j)$ is given by

$$R(i_0,j_0)^{-1} = \inf\{Q(f);\ f(i_0) - f(j_0) = 1\} . \tag{13}$$

We also give a second definition of resistance

$$R_1(i_0,j_0)^{-1} = \inf\{Q(f);\ f(i_0) - f(j_0) = 1 \quad f \in \mathcal{D}_0\}$$

where $\mathcal{D}_0$ is the set of the functions on $\mathcal{L}$ with finite support. For all finite graphs and most graphs of physical interest $R_1(i,j) = R(ij)$. In fact, if $R_1(i_0,j_0) < R(i_0,j_0)$ for some $i_0,j_0 \in \mathcal{L}$ then there must exist a non-constant bounded harmonic function on $\mathcal{L}$.

Let $\mathcal{L} = \mathbb{Z}^n$ for $n = 1,2,\ldots$ and let G be the graph obtained by connecting nearest neighbors of $\mathcal{L}$ with one ohm resistors. For these graphs $R_1(ij) = R(ij)$ for all $i,j \in \mathcal{L}$. For $n = 1$ the resistance $R(ij) = |i - j|$, the distance between i and j. For $n = 2$ the resistance $R(ij)$ grows logarithmically with the distance between i and j. For $n = 3$ the resistance $R(ij)$ is bounded, in fact

$$R(ij) \to .50545 \quad \text{as} \quad |i - j| \to \infty .$$

We now state some resistance inequalities for the Heisenberg model.

THEOREM 1. Suppose ω is a state of the Heisenberg spin algebra $\mathfrak{A}$ and H is a Hamiltonian given in (2). Then

$$\omega(1-(\vec{\sigma}_{i_0} \cdot \vec{\sigma}_{j_0})) \leq R(i_0,j_0)\omega(H) = R(i_0,j_0)\ \Sigma_{(ij)\in G} J(ij)\ \omega(I - \vec{\sigma}_i \cdot \vec{\sigma}_j)$$

where $R(i_0, j_0)$ is the resistance between i_0 and j_0.

COROLLARY. Suppose ω is a state of the Heisenberg model in three dimensions with nearest neighbor interaction of unit strength. Then if ω is a state of finite energy then for every ϵ, there is a finite set S so that

$$\omega(\vec{\sigma}_i \cdot \vec{\sigma}_j) \geq 1 - \epsilon$$

for all $i, j \notin S$.

THEOREM 2. Suppose ω is an (α_t, β)-KMS state of the Heisenberg spin algebra $\mathfrak{A}$ associated with a Hamiltonian given by equation (2). Then

$$|\omega(\vec{\sigma}_{i_0} \cdot \vec{\sigma}_{j_0})|^2 \leq \frac{6\beta}{R_1(i_0, j_0)}$$

where $R_1(i_0, j_0)$ is the resistance given by equation (14).

This result generalizes the well known theorem of Mermin and Wagner [2]. It follows from this theorem or was shown by Mermin and Wagner that one and 2-dimensional Heisenberg models do not have long range order since in one and 2-dimensions $R(ij)$ grows without bound on $|i - j| \to \infty$.

We conclude with a conjecture.

CONJECTURE. There is a constant K_0 (independent of G) so that if ω is an (α_t, β)-KMS state of $\mathfrak{A}$ then

$$\omega(1 - \vec{\sigma}_i \cdot \vec{\sigma}_j) \leq K_0 \, \beta^{-1} \, R_1'(i, j)$$

where

$$R_1'(i_0, j_0) = \inf\{\Sigma_{ij \in G} \, J(ij)\omega(\vec{\sigma}_i \cdot \vec{\sigma}_j)(f(i) - f(j))^2 : f \in \mathcal{D}_0, \; f(i_0) - f(j_0) = 1\}$$

The paper of Frohlich, Simon and Spencer when combined with the resistance calculations described here shows this conjecture is true for the Heisenberg model in n-dimensions with unit nearest neighbor interaction with the constant $K_0 = 3/2$.

The truth of this conjecture would establish the existence of long range order for a large class of physical models. The truth of this conjecture plus Theorem 2 would show that the question of long range order for the isotropic Heisemberg model could be determined from resistance calculations.

REFERENCES

1. Fröhlich, Simon and Spencer, preprint.
2. Mermin and Wagner, Phys. Rev. Letters 17, 1133 (1966).
3. R. Powers, Jour. of Math. Phys. 17, 1910 (1976).
4. R. Powers, Comm. of Math. Phys. 51, 151 (1976).

Homogeneity of the state space of factors of type III_1

Erling Størmer

1. Introduction. In this note I'll describe some work of A. Connes and myself on the state space of III_1- factors [3]. The main result states that for such factors the orbit of each normal state under the action of inner automorphisms is norm dense in the set of all normal states. This answers to the affirmative a conjecture of Connes and Takesaki [4]. Our main technical tool is a generalization of the skew information in factors of type I introduced by Wigner and Yanase [7].

2. Skew informations. Let M be a von Neumann algebra with separable predual. Let $\varphi \in M_*^+$, and assume for simplicity that φ is the vector state ω_{ξ_φ} defined by a separating and cyclic vector ξ_φ (this is unnecessary, use the cone $P^\natural$). Let J be the involution defined by M and ξ_φ, and σ^φ the modular automorphism corresponding to φ. The skew information is defined by

$$I(\varphi,x) = \tfrac{1}{2}\|(Jx^*J - x)\xi_\varphi\|^2, \quad x \in M .$$

Note that x belongs to the centralizer M_φ of φ in M if and only if $I(\varphi,x) = 0$. We note two of the most important properties of $I(\varphi,x)$.

Lemma 1. Let φ be as above. Then

(i) $\|[\varphi,x]\| \leq 2^{3/2}\varphi(1)^{1/2} I(\varphi,x)^{1/2}, \quad x \in M.$

(ii) If $x \in M(\sigma^\varphi,[(1-\delta)^2,(1+\delta)^2])$, where the spectral subspace is taken with respect to the multiplicative group of positive reals, then $I(\varphi,x) \leq \tfrac{1}{2}\delta^2(\|x\|_\varphi)^2$.

Here and later $(\|x\|_\varphi)^2 = \varphi(x^*x)$, and $(\|x\|_\varphi^\#)^2 = \varphi(x^*x + xx^*)$. Note that (i) says that if $I(\varphi,x)$ is small, then x almost commutes with φ, and (ii) says that this holds for all x in a spectral subspace corresponding to a small neighborhood of 1.

We next show that we can find partial isometries with small skew information. For this we follow [2] and let E_a be the characteristic function of the interval $[a,+\infty)$ in $\mathbb{R}$, $a > 0$. For $x \in M$ with polar decomposition $x = u(x)|x|$ we let $u_a(x) = u(x)E_a(|x|)$. Modifying the techniques in [2] we can show

Lemma 2. If $\varphi \in M_*^+$ and $x \in M$ we have

(i) $\displaystyle\int_0^\infty (\|u_{a^{\frac{1}{2}}}(x)\|_\varphi^\#)^2\, da = (\|x\|_\varphi^\#)^2$

(ii) $\displaystyle\int_0^\infty I(\varphi, u_{a^{\frac{1}{2}}}(x))\, da \le 6\, I(\varphi,x)^{\frac{1}{2}} \|x\|_\varphi^\#$.

In particular if $I(\varphi,x) \le \epsilon(\|x\|_\varphi^\#)^2$ then

$$\int_0^\infty I(\varphi, u_{a^{\frac{1}{2}}}(x))\, da \le 6\,\epsilon^{\frac{1}{2}}(\|x\|_\varphi^\#)^2 = 6\,\epsilon^{\frac{1}{2}} \int_0^\infty (\|u_{a^{\frac{1}{2}}}(x)\|_\varphi^\#)^2\, da .$$

Since this holds for all $a > 0$ there is a such that $u_a(x) \neq 0$ and

(1) $\qquad I(\varphi, u_a(x)) \le 7\,\epsilon^{\frac{1}{2}}(\|u_a(x)\|_\varphi^\#)^2$.

Lemma 3. Let $\varphi \in M_*^+$, where M is a factor of type III_1. Let $e', f' \in M_\varphi$ be nonzero projections smaller than the support of φ. For any $\epsilon > 0$ there exists a partial isometry $u \neq 0$ in M such that $u^*u = e \le e'$, $uu^* = f \le f'$, and

(i) $I(\varphi,u) \le \epsilon(\|u\|_\varphi^\#)^2$.

(ii) $I(\varphi,e) \le \epsilon\,\varphi(e)$, $I(\varphi,f) \le \epsilon\,\varphi(f)$.

Since M is a factor of type III_1 (this is where we use this assumption) there is $x = f'xe' \neq 0$ in $M(\sigma^{\varphi},[(1-\delta),(1+\delta)])$ for each $\delta > 0$. By (1) above and Lemma 1 (ii) we can find a partial isometry $u \in f'Me'$ such that $I(\varphi,u) \leq \epsilon(\|u\|_{\varphi}^{\#})^2$. This shows (i), and (ii) is not hard.

3. The main theorem.

Theorem. Let M be a factor of type III_1 with separable predual. Then for any $\epsilon > 0$ and normal states φ and ψ there exists a unitary u in M such that $\|\varphi_u - \psi\| < \epsilon$, where $\varphi_u(x) = \varphi(u^* x u)$, $x \in M$.

The proof is a rather complicated transfinite induction argument based mainly on Lemma 3. We consider the (nonnormalized) state $\theta = \begin{pmatrix} \varphi & 0 \\ 0 & \psi \end{pmatrix}$ on $M \otimes M_2(\mathbb{C})$ defined by $\theta((x_{ij})) = \varphi(x_{11}) + \psi(x_{22})$, and try to find a unitary $\bar{u} = \begin{pmatrix} 0 & 0 \\ u & 0 \end{pmatrix}$ approximately in the centralizer of θ. The main idea is to use Lemma 3 in order to build up such a unitary as a sum of partial isometries with mutually orthogonal supports and ranges. In the lemma we let $e' \leq \begin{pmatrix} 1 & \\ & 0 \end{pmatrix}$ and $f' \leq \begin{pmatrix} 0 & \\ & 1 \end{pmatrix}$ so the u obtained is always of the form $\bar{u}$ above.

I next indicate some consequences of the theorem. Recall that for $\lambda \in [0,\frac{1}{2}]$ M has property L_λ of Powers [6] if for $\varphi \in M_*^+$ and $\epsilon > 0$ there exists a partial isometry $u \in M$ such that $u^2 = 0$, $u^*u + uu^* = 1$ and $|\lambda\varphi(ux) - (1-\lambda)\varphi(xu)| \leq \epsilon\|x\|$ for $x \in M$.

Corollary 1. Let $\lambda \in [0,\frac{1}{2})$. Then a factor M has property L_λ if and only if $1/(1-\lambda) \in S(M)$.

From [1, 6] all that remains is to show that if M is of type III_1 then M has property L_λ. For this use a state of the form $\varphi \otimes \omega_\lambda$ on $M \otimes M_2(\mathbb{C})$ $(\cong M)$, where $\omega_\lambda((x_{ij})) = \lambda x_{11} + (1-\lambda)x_{22}$. It is clear that $\varphi \otimes \omega_\lambda$ satisfies the property of L_λ (also for $\lambda = \frac{1}{2}$), hence all states do by the theorem.

Note that the above argument shows that the assertion of the theorem characterizes III_1-factors. Another immediate consequence of the theorem is

Corollary 2. Let $(M_\nu)_{\nu \in A}$ be a denumerable family of factors of type III_1. Then the infinite tensor product $\underset{\nu}{\otimes} (M_\nu, \varphi_\nu)$ is up to isomorphism, independent of the choice of the sequence $(\varphi_\nu), \varphi_\nu$ normal state of M_ν.

Corollary 3. Let R be the hyperfinite II_1-factor and M a non type I factor. Then there exists a faithful normal state on M whose centralizer contains R.

By [1] we may assume M is of type III_1, so M has property $L_{\frac{1}{2}}$ from the proof of Corollary 1. Then we can, using the Theorem, find a sequence of states like $\varphi \otimes \omega_{\frac{1}{2}} \otimes \dots \otimes \omega_{\frac{1}{2}}$ which converges in norm to the desired state.

Dell' Antonio [5] said a factor M has property U if each sequence of normal states on M which converges weakly to a normal state already converges uniformly. He showed that type I factors have property U and conjectured the converse. Since he also showed that R does not have property U it is easy from Corollary 3 to obtain

Corollary 4. A factor has property U if and only if it is of type I.

References

1. A. Connes, Une classification des facteurs de type III, Ann. Scient. École Norm. Sup. 4. série 6 (1973), 133-252.

2. ———, Classification of injective factors, Annals of Math. 104 (1976), 73-115.

3. A. Connes and E. Størmer, Homogeneity of the state space of factors of type III_1, J. Functional Anal. To appear.

4. A. Connes and M. Takesaki, The flow of weights on factors of type III. To appear.

5. G.F. Dell'Antonio, On the limit of sequences of normal states, Comm. pure appl. math. 20 (1967), 413-429.

6. R.T. Powers, UHF-algebras and their applications to representations of the anticommutation relations, Cargèse Lectures in Physics, Vol. 4, Gordon and Breach, New York - London - Paris 1970, 137-168.

7. E.P. Wigner and M.M. Yanase, Information contents of distributions, Proc. Nat. Acad. Sciences, 49 (1963), 910-918.

PRODUCT ISOMETRIES AND AUTOMORPHISMS OF THE CAR ALGEBRA

by

Richard V Kadison*

I. INTRODUCTION.

The methods of multilinear algebra and, in particular, those of the exterior calculus provide a useful framework for studying the Fock space, $\mathcal{H}_{\mathcal{F}}$, of antisymmetrized wave functions and the Fock representation of the Canonical Anticommutation Relations (CAR) on it. With the aid of these methods, we study the structure of certain mappings, product isometries, of the n-particle subspace $\mathcal{H}_n$ of $\mathcal{H}_{\mathcal{F}}$.

With $\mathcal{H}$ a complex Hilbert space, we denote by $f_1 \wedge \dots \wedge f_n$ the vector $(n!)^{-\frac{1}{2}}\Sigma_\sigma \chi(\sigma) f_{\sigma(1)} \otimes \dots \otimes f_{\sigma(n)}$ in $\mathcal{H} \otimes \dots \otimes \mathcal{H}$, the n-fold tensor product of $\mathcal{H}$ with itself, where σ is a permutation of $\{1,\dots,n\}$ and $\chi(\sigma)$ is its sign. The space spanned by $\{f_1 \wedge \dots \wedge f_n: f_j \in \mathcal{H}\}$ is denoted by $\mathcal{H}_n$; and $\mathcal{H}_{\mathcal{F}}$, (antisymmetric) Fock space, is $\Sigma_{n=0}^{\infty} \oplus \mathcal{H}_n$, where $\mathcal{H}_1$ is $\mathcal{H}$ and $\mathcal{H}_0$ is a 1-dimensional space generated by a unit vector, Φ_0, the Fock vacuum. A linear isometry V of some subspace $\mathcal{K}_n$ $(=\{f_1 \wedge \dots \wedge f_n: f_j \in \mathcal{K} \subseteq \mathcal{H}\})$ is said to be a product isometry when $V(f_1 \wedge \dots \wedge f_n)$ is a product vector (i.e. of the form $g_1 \wedge \dots \wedge g_m$) in $\mathcal{H}_{\mathcal{F}}$. One of our principal aims is the following result.

* With partial support of NSF.

THEOREM A. If $\hat{U}$ is a product unitary transformation of $\mathcal{H}_n$ onto $\mathcal{H}_n$, there is a unitary transformation U of $\mathcal{H}$ onto $\mathcal{H}$ such that $\hat{U}(f_1 \wedge \dots \wedge f_n) = Uf_1 \wedge \dots \wedge Uf_n$.

This is proved by a combinatorial-geometric study of the way in which $\hat{U}$ transforms the subspace $[f_1,\dots,f_n]$ of $\mathcal{H}$ associated with $f_1 \wedge \dots \wedge f_n$. We note that

$$\langle f_1 \wedge \dots \wedge f_n | g_1 \wedge \dots \wedge g_n \rangle = \det\left(\langle f_i | g_j \rangle\right) ,$$

and that $af_1 \wedge \dots \wedge f_n = g_1 \wedge \dots \wedge g_n \neq 0$ if and only if the spaces $[f_1 \wedge \dots \wedge f_n]$ $(=[f_1, \dots, f_n])$ and $[g_1, \dots, g_n]$ associated with $f_1 \wedge \dots \wedge f_n$ and $g_1 \wedge \dots \wedge g_n$ are n-dimensional and coincide.

Applying Theorem A, one can, then, show that:

THEOREM B. If α is an automorphism of the CAR algebra $\mathfrak{A}$ whose transpose $\hat{\alpha}$ maps the set $\mathfrak{G}$ of pure, gauge-invariant, quasi-free states of $\mathfrak{A}$ onto itself, then there is a unitary operator U on $\mathcal{H}$ such that $\alpha(a(f)) = a(Uf)$, for all f in $\mathcal{H}$; or there is a conjugate-linear, unitary operator W on $\mathcal{H}$ such that $\alpha(a(f)) = a(Wf)^*$, for all f in $\mathcal{H}$, where a(f) is the annihilator for a particle with wave function f.

Note that one can read from this result the fact that $\hat{\alpha}$ transforms the Fock (vacuum) state φ_0 either onto itself or onto the anti-Fock state φ_I; though this fact is established as a preliminary to proving Theorem B.

The <u>creator</u> $a(f)^*$, determined by :

$$a(f)^*(f_1 \wedge \dots \wedge f_n) = f \wedge f_1 \wedge \dots \wedge f_n ,$$

is the adjoint of the <u>annihilator</u> a(f) determined by:

$$a(f)(f_1 \wedge \dots \wedge f_n) = \sum_{j=1}^{n} (-1)^{j+1} \langle f | f_j \rangle \, f_1 \wedge \dots \wedge f_{j-1} \wedge f_{j+1} \wedge \dots \wedge f_n .$$

The mapping $f \to a(f)$ is conjugate linear (the inner product on $\mathcal{H}$ being linear in its second argument) and satisfies $a(f)a(g) + a(g)a(f) = 0$, $a(f)a(g)^* + a(g)^*a(f) = \langle f|g\rangle I$ (the CAR). A conjugate-linear mapping of $\mathcal{H}$ onto operators $a(f)$ on a Hilbert space $\mathcal{K}$ satisfying the CAR is said to be a representation of the CAR (over $\mathcal{H}$ on $\mathcal{K}$). The representing operators $a(f)$ are partial isometries (with initial and final spaces orthogonal having sum $\mathcal{K}$), hence, bounded. The particular representation of the CAR on $\mathcal{H}_{\mathfrak{F}}$ which we have described is the Fock representation. The C*-algebra $\mathfrak{A}$, generated by $\{a(f), a(f)^* : f \in \mathcal{H}\}$, is the CAR algebra. Its representations are in one-one correspondence with those of the CAR.

If A is an operator on $\mathcal{H}$ such that $0 \le A \le I$, defining $\varphi_A(a(f_n)^* \dots a(f_1)^*a(g_1) \dots a(g_n)^*)$ to be $\det(\langle g_i|Af_j\rangle)$ $(= \langle g_1 \wedge \dots \wedge g_n|Af_1 \wedge \dots \wedge Af_n\rangle)$ determines a state φ_A of $\mathfrak{A}$, the gauge-invariant, quasi-free state (with one-particle operator A). The state φ_E is pure if and only if E is a projection on $\mathcal{H}$; and φ_E is equivalent to the Fock state φ_0 if and only if E is a projection with finite-dimensional range. In case $E(\mathcal{H})$ is finite-dimensional and $\{e_1, \dots, e_n\}$ is an orthonormal basis for it, we have $\varphi_E(A) = \langle e_1 \wedge \dots \wedge e_n|A(e_1 \wedge \dots \wedge e_n)\rangle$ $= \omega_{e_1 \wedge \dots \wedge e_n}(A)$, for each A in $\mathfrak{A}$. The intimate relation between Theorems A and B is a consequence of these last comments, for, then, a unitary operator on $\mathcal{H}_{\mathfrak{F}}$ implementing α transforms product vectors into product vectors.

The work on which these notes are based is a joint study conducted with N M Hugenholtz. An extended account is to be found in: "Automorphisms and Quasi-free States of the CAR Algebra", Commun. Math. Phys. 43 (1975), pp 181-197. The author acknowledges with gratitude the support of the SRC and the hospitality of the Department of Mathematics at the University of Newcastle, as well as the support and hospitality of the Centre Universitaire de Luminy-Marseille, where parts of this work were completed.

II. PRODUCT UNITARIES.

If V is a product isometry of an infinite-dimensional subspace $\mathcal{K}$ of $\mathcal{H}$ $(=\mathcal{H}_1)$ into $\mathcal{H}_m$, then, with $\{e_j\}$ an orthonormal basis for $\mathcal{K}$, the fact that Ve_j and Ve_k are orthogonal product vectors in $\mathcal{H}_m$ and $Ve_j + Ve_k$ $(= V(e_j+e_k))$ is also a product vector in $\mathcal{H}_m$ leads to the conclusion that the projections E_j and E_k with the m-dimensional ranges $[Ve_j]$ and $[Ve_k]$ commute and $[Ve_j] \cap [Ve_k]$ has dimension m-1. It follows that $\bigcap_j E_j(\mathcal{H})$ has dimension m-1; and, hence, $\cap[Vx]$ has dimension m-1.

If W is a product isometry of $\mathcal{K}_n$ into $\mathcal{H}_{\mathcal{F}}$, isometry considerations show that W has range in one $\mathcal{H}_m$. If $\mathcal{K}$ is infinite dimensional and $n \leq m$ then $[W(f_1 \wedge \cdots \wedge f_n)] \cap [W(g_1 \wedge \cdots \wedge g_n)]$ has dimension at least m-n. To see this, we may assume that each of $\{f_1,\ldots,f_n\}$ and $\{g_1,\ldots,g_n\}$ are orthonormal sets and make use of the mapping

$f \to W(f \wedge f_2 \wedge \dots \wedge f_n)$, which is a product isometry of $\mathcal{K} \ominus [f_2, \dots, f_n]$ into $\mathcal{H}_m$. From the preceding, it follows that $[W(h_1 \wedge f_2 \wedge \dots \wedge f_n)]$ and $[W(f_1 \wedge \dots \wedge f_n)]$ have an intersection of dimension at least m-1, where h_1 is a vector in $[g_1, \dots, g_n]$ orthogonal to $[f_2 \wedge \dots \wedge f_n]$. In the same way, choosing h_2 in $[g_1, \dots, g_n]$ orthogonal to $[h_1, f_3, \dots, f_n]$, we see that $[W(h_1 \wedge h_2 \wedge f_3 \wedge \dots \wedge f_n)] \cap [W(f_1 \wedge \dots \wedge f_n)]$ has dimension at least m-2. Continuing, we have that $[W(h_1 \wedge \dots \wedge h_n)] \cap [W(f_1 \wedge \dots \wedge f_n)]$ has dimension at least m-n, and $h_1 \wedge \dots \wedge h_n = cg_1 \wedge \dots \wedge g_n$.

If $\{e_j\}$ is an orthonormal basis for $\mathcal{K}$ and $i_1, \dots, i_n, j_i, \dots, j_n$ is such that $[W(e_{i_1} \wedge \dots \wedge e_{i_n})]$ and $[W(e_{j_1} \wedge \dots \wedge e_{j_n})]$ have intersection of dimension precisely m-n then $\bigcap_{x_1, \dots, x_n} [W(x_1 \wedge \dots \wedge x_n)]$ has dimension m-n. This amounts to showing that

$$[W(e_{i_1} \wedge \dots \wedge e_{i_n})] \cap [W(e_{j_1} \wedge \dots \wedge e_{j_n})] \subseteq [W(e_{k_1} \wedge \dots \wedge e_{k_n})] \qquad (*)$$

for all $k_1, \dots, k_n$. This is effected by arguing inductively on n - the conclusion of the preceding argument allowing us to carry the hypothesis of an intersection having dimension m-n to one where the intersection has dimension m-(n-1). For this purpose, we use the mapping

$$x_1 \wedge \dots \wedge x_{r-1} \wedge x_{r+1} \wedge \dots \wedge x_n \to W(x_1 \wedge \dots \wedge x_{r-1} \wedge e_{i_r} \wedge x_{r+1} \wedge \dots \wedge x_n)$$

of $(\mathcal{K} \ominus [e_{i_r}])_{n-1}$ into $\mathcal{H}_m$. As an intermediate conclusion, we obtain (*) if at least one of $k_1, \dots, k_n$ is in $\{i_1, \dots, i_n, j_1, \dots, j_n\}$. If we know that the intersection, M, of $[W(e_{i_1} \wedge \dots \wedge e_{i_n})]$ and $[W(e_{j_1} \wedge \dots \wedge e_{j_{r-1}} \wedge e_t \wedge e_{j_{r+1}} \wedge \dots \wedge e_{j_n})]$ has dimension m-n, when $t \notin \{i_1, \dots, i_n\}$; then, if $t = k_r \notin \{i_1, \dots, i_n\}$, we have, from our intermediate conclusion, that $[W(e_{k_1} \wedge \dots \wedge e_{k_n})]$ contains M while M contains $[W(e_{i_1} \wedge \dots \wedge e_{i_n})] \cap [W(e_{j_1} \wedge \dots \wedge e_{j_n})]$.

To see that M has dimension m-n, note that $\bigcap_t [W(e_{j_1} \wedge \cdots \wedge e_{j_{r-1}} \wedge e_t \wedge e_{j_{r+1}} \wedge \cdots \wedge e_{j_n})]$ (=N) has dimension m-1, from our initial observation. Thus each $[W(e_{j_1} \wedge \cdots \wedge e_{j_{r-1}} \wedge e_t \wedge e_{j_{r+1}} \wedge \cdots \wedge e_{j_n})]$ is generated by N and a vector g_t orthogonal to it. Since $\{W(e_{j_1} \wedge \cdots \wedge e_{j_{r-1}} \wedge e_t \wedge e_{j_{r+1}} \wedge \cdots \wedge e_{j_n})\}$ is a family of orthogonal vectors, $\{g_t\}$ is such a family; and no $[W(e_{j_1} \wedge \cdots \wedge e_{j_{r-1}} \wedge e_t \wedge e_{j_{r+1}} \wedge \cdots \wedge e_{j_n})]$ is contained in a union of the others. If M has dimension greater than m-n it has a vector orthogonal to $[W(e_{i_1} \wedge \cdots \wedge e_{i_n})] \cap [W(e_{j_1} \wedge \cdots \wedge e_{j_n})]$ as do the intersections of $[W(e_{i_1} \wedge \cdots \wedge e_{i_n})]$ and $[W(e_{j_1} \wedge \cdots \wedge e_{j_{r-1}} \wedge e_{i_s} \wedge e_{j_{r+1}} \wedge \cdots \wedge e_{j_n})]$. Each of these n+1 vectors (taking s to be 1, ..., n) are not in the m 1 dimensional space N, for, otherwise, they are in $[W(e_{j_1} \wedge \cdots \wedge e_{j_n})]$, hence in $[W(e_{i_1} \wedge \cdots \wedge e_{i_n})] \cap [W(e_{j_1} \wedge \cdots \wedge e_{j_n})]$, contrary to choice. Thus each of these vectors generates with N its corresponding m-dimensional space $[W(e_{j_1} \wedge \cdots \wedge e_{j_{r-1}} \wedge e_{i_s} \wedge e_{j_{r+1}} \wedge \cdots \wedge e_{j_n})]$ (or $[W(e_{j_1} \wedge \cdots \wedge e_{j_{r-1}} \wedge e_{k_n} \wedge e_{j_{r+1}} \wedge \cdots \wedge e_{j_n})]$). A linear relation among these vectors would imply, therefore, that one of these spaces is contained in the union of the others - contrary to what we have noted. But these n+1 vectors and the m-n-dimensional space $[W(e_{i_1} \wedge \cdots \wedge e_{i_n})] \cap [W(e_{j_1} \wedge \cdots \wedge e_{j_n})]$ orthogonal to them are all contained in $[W(e_{i_1} \wedge \cdots \wedge e_{i_n})]$, an m dimensional space. Thus there must be a linear relation among these n+1 vectors. From this contradiction, we conclude that M has dimension m-n.

Summarizing, to this point, we have proved:

Proposition C. If W is a product isometry of $\mathcal{K}_n$ into $\mathcal{H}_m$, where $n \le m$ and $\mathcal{K}$ is an infinite-dimensional subspace of $\mathcal{H}$, and the intersection of $[W(e_{i_1} \wedge \cdots \wedge e_{i_n})]$ and $[W(e_{j_1} \wedge \cdots \wedge e_{j_n})]$ has dimension $m-n$, for some $e_{i_1}, \ldots, e_{i_n}, e_{j_1}, \ldots, e_{j_n}$, then $\bigcap_{x_1,\ldots,x_n} [W(x_1 \wedge \cdots \wedge x_n)]$ has dimension $m-n$.

It follows, without difficulty, from this proposition, that a product unitary defined on $\mathcal{H}_{\mathcal{F}}$ maps each $\mathcal{H}_n$ onto $\mathcal{H}_n$. If $\hat{U}$ is a product unitary on $\mathcal{H}_n$ then $[x_1, \ldots, x_n] \cap \ldots \cap [z_1, \ldots, z_n]$ and $[\hat{U}(x_1 \wedge \cdots \wedge x_n)] \cap \ldots \cap [\hat{U}(z_1 \wedge \cdots \wedge z_n)]$ have the same dimension (for infinite, as well as, finite intersections). In particular, $\bigcap_{x_1,\ldots,x_{n-1}} [\hat{U}(x_1 \wedge \cdots \wedge x_{n-1} \wedge e)]$ has dimension 1 for each unit vector e in $\mathcal{H}$. If $\{e_j\}$ is an orthonormal basis for $\mathcal{H}$ and f_j' is a unit vector in $\bigcap[\hat{U}(x_1 \wedge \cdots \wedge x_{n-1} \wedge e_j)]$ then $\{f_j'\}$ is an orthonormal basis for $\mathcal{H}$. To see this, we note $f_j' \in M_k$ when $k \ne j$, where $M_k = [\hat{U}(e_1 \wedge \cdots \wedge e_{k-1} \wedge e_{k+1} \wedge \cdots \wedge e_{n+1})]$; and f_j' is orthogonal to M_j. From Proposition C and the consequences noted following it, $M_1 \vee M_2$ is an $n+1$-dimensional space containing each (n-dimensional) M_j, $j = 1, \ldots, n+1$. Thus $f_1', \ldots, f_{n+1}'$ is an orthonormal basis for $M_1 \vee M_2$.

It follows, now, that $\hat{U}(e_{i_1} \wedge \cdots \wedge e_{i_n}) = c_{i_1 \ldots i_n} f_{i_1}' \wedge \cdots \wedge f_{i_n}'$; where $|c_{i_1 \ldots i_n}| = 1$. Writing c_j' for $c_{1\ldots j-1 j+1\ldots n+1}$, c_j for $\overline{c_j'} \prod_{k=1}^{n+1} c_k'$, and f_j for $c_j f_j'$, we have

$$\hat{U}(e_1 \wedge \cdots \wedge e_{j-1} \wedge e_{j+1} \wedge \cdots \wedge e_{n+1}) = f_1 \wedge \cdots \wedge f_{j-1} \wedge f_{j+1} \wedge \cdots \wedge f_{n+1},$$

for $j = 1, \ldots, n+1$. Using the fact that $\hat{U}$ is a product unitary on $\mathcal{H}_n$, it follows that $\hat{U}(e_1 \wedge \cdots \wedge e_{j-1} \wedge e_{j+1} \wedge \cdots \wedge e_{n+1} \wedge e_{n+2}) = cf_1 \wedge \cdots \wedge f_{j-1} \wedge f_{j+1} \wedge \cdots \wedge f_{n+1} \wedge f'_{n+2}$, where the phase factor c is the same for $j = 1, \ldots, n+1$. Taking f_{n+2} to be cf'_{n+2}, we construct, inductively, an orthonormal basis $\{f_j\}$ for $\mathcal{H}$ such that $\hat{U}(e_{i_1} \wedge \cdots \wedge e_{i_n}) = f_{i_1} \wedge \cdots \wedge f_{i_n}$ for all $i_1, \ldots, i_n$. Theorem A results from letting U be the unitary operator on $\mathcal{H}$ determined by: $Ue_j = f_j$.

III. THE AUTOMORPHISM.

Suppose, now, that $\hat{U}$ is a product unitary on $\mathcal{H}_{\mathcal{F}}$ that induces an automorphism of $\mathfrak{A}$. Then $\hat{U}$ maps $\mathcal{H}_n$ onto $\mathcal{H}_n$ for each n; and there is a unitary operator U_n on $\mathcal{H}$ such that $\hat{U}(x_1 \wedge \cdots \wedge x_n) = U_n x_1 \wedge \cdots \wedge U_n x_n$. It follows that $\hat{U}a(f)a(f)^*\hat{U}^*$ and $a(U_n f)a(U_n f)^*$ have the same restriction to $\mathcal{H}_n$, for each f in $\mathcal{H}$. A calculation (see Appendix I) shows that $U_n f = c_{fnm} U_m f$, for all f, n and m, in this case. An (easy) algebraic lemma (see Appendix III) allows us to conclude that $U_n = c_{nm} U_m$. Hence,

$$\hat{U}a(f)^*\hat{U}^*(g_1 \wedge \cdots \wedge g_n) = \hat{U}(f \wedge U_n^* g_1 \wedge \cdots \wedge U_n^* g_n) = U_m f \wedge U_m U_n^* g_1 \wedge \cdots \wedge U_m U_n^* g_n$$

$$= c_{mn}^n c_{m1} U_1 f \wedge g_1 \wedge \cdots \wedge g_n = c_n a(U_1 f)^*(g_1 \wedge \cdots \wedge g_n)$$

so that $\hat{U}a(f)^*\hat{U}^*$ and $c_n a(U_1 f)^*$ have the same restriction to $\mathcal{H}_n$. Another calculation (see Appendix II) shows that $\hat{U}a(f)^*\hat{U}^* = c_f a(U_1 f)^*$. Applying our algebraic lemma, again, $\hat{U}a(f)^*\hat{U}^* = ca(U_1 f)^*$ (on $\mathcal{H}_{\mathcal{F}}$, the phase factor c is no longer dependent on f). Finally, as

$$\hat{U}a(f)^*\hat{U}^*\Phi_0 = \hat{U}f = U_1 f = a(U_1 f)^*\Phi_0 ,$$

$c = 1$; and $\hat{U}a(f)^*\hat{U}^* = a(U_1 f)^*$, for each f in $\mathcal{H}$.

Writing U for U_1, it follows that

$\hat{U}a(f_1)^* \dots a(f_n)^*\hat{U}^*\Phi_0 = \hat{U}(f_1 \wedge \dots \wedge f_n) = a(Uf_1)^* \dots a(Uf_n)^*\Phi_0 = Uf_1 \wedge \dots \wedge Uf_n$,

for all $f_1, \dots, f_n$ in $\mathcal{H}$ and all n.

If α is an automorphism of $\mathfrak{A}$ such that $\hat{\alpha}(\mathfrak{S}) = \mathfrak{S}$ and $\hat{\alpha}(\varphi_0) = \varphi_0$, then α is implemented by a unitary operator $\hat{U}$ on $\mathcal{H}_{\mathcal{F}}$. Since the states in $\mathfrak{S}$ equivalent to φ_0 are precisely those vector states of $\mathfrak{A}$ corresponding to product vectors;

$$\hat{\alpha}(\omega_{f_1 \wedge \dots \wedge f_n}) = \omega_{\hat{U}(f_1 \wedge \dots \wedge f_n)}|\mathfrak{A} = \omega_{g_1 \wedge \dots \wedge g_n}|\mathfrak{A} .$$

Since $\mathfrak{A}$ acts irreducibly on $\mathcal{H}_{\mathcal{F}}$; we conclude that $\hat{U}(f_1 \wedge \dots \wedge f_n)$ is a product vector (a scalar multiple of $g_1 \wedge \dots \wedge g_n$). From the preceding section, $\hat{U}a(f)\hat{U}^* = a(Uf)$, where U (a unitary operator on $\mathcal{H}$) is the restriction of $\hat{U}$ to $\mathcal{H}$.

If $\hat{\alpha}(\varphi_0) = \varphi_I$, V is a conjugate linear unitary operator on $\mathcal{H}$, and σ is the automorphism of $\mathfrak{A}$ determined by $\sigma(a(f)) = a(Vf)^*$, then $\widehat{\alpha\circ\sigma}(\varphi_0) = \varphi_0$ and $\widehat{\alpha\circ\sigma}(\mathfrak{S}) = \mathfrak{S}$. Hence there is a unitary operator U on $\mathcal{H}$ such that

$$a(Wf)^* = a(UV^{-1}f)^* = (\alpha\circ\sigma)(a(V^{-1}f)^*) = \alpha(a(f)) ,$$

where W is the conjugate-linear unitary operator UV^{-1} on $\mathcal{H}$.

Theorem B follows once we show that $\hat{\alpha}$ maps φ_0 onto φ_0 or φ_I. In any event, $\hat{\alpha}(\varphi_0) \in \mathfrak{S}$; so that $\hat{\alpha}(\varphi_0) = \varphi_{E_1}$ for some projection E_1 on $\mathcal{H}$. We derive a contradiction from the assumption that E_1 is neither 0 nor I. With this assumption, there are unit vectors f_1 and f_2 in and orthogonal to $E_1(\mathcal{H})$. If $f_3 = \frac{1}{\sqrt{2}}(f_1+f_2)$ and E_0, E_2, E_3 are the projections with ranges $E_1(\mathcal{H}) \ominus [f_1]$, $E_0(\mathcal{H}) \oplus [f_2]$, $E_0(\mathcal{H}) \oplus [f_3]$, then $\varphi_{E_0}(A) = \varphi_{E_j}(a(f_j)^*Aa(f_j))$, $j = 1, 2, 3$. Since the φ_{E_j} are equivalent, the states $\hat{\alpha}^{-1}(\varphi_{E_j})$ are in $\mathfrak{S}$ and equivalent to $\varphi_0(=\hat{\alpha}^{-1}(\varphi_{E_1}))$.

Thus $\hat{\alpha}^{-1}(\varphi_{E_j}) = \omega_{g_j}|\mathfrak{A}$, where g_j is a unit product vector in $\mathcal{H}_{\mathcal{F}}$. Now,

$$\varphi_{E_j}(a(f_j)^*\alpha^{-1}(A)a(f_j)) = \varphi_{E_j}(\alpha^{-1}(\alpha(a(f_j)^*)A\alpha(a(f_j))) = \omega_{\alpha(a(f_j))g_j}(A)$$
$$= \hat{\alpha}^{-1}(\varphi_{E_0})(A) = \psi_{g_0}(A) ,$$

for $j = 1, 2, 3$. Since $\mathfrak{A}$ acts irreducibly on $\mathcal{H}_{\mathcal{F}}$, $\alpha(a(f_j))g_j = c_j g_0$, where $|c_j| = 1$. As $a(f_j)$ is a partial isometry and $\|\alpha(a(f_j))g_j\| = \|c_j g_0\| = 1$; $\alpha(a(f_j))$ is a partial isometry with g_j in its initial space and g_0 in its final space. Thus $g_3 = \frac{c_3}{\sqrt{2}}[\alpha(a(f_1)^*)g_0 + \alpha(a(f_2)^*)g_0] = \frac{c_3}{\sqrt{2}}[\bar{c}_1 g_1 + \bar{c}_2 g_2]$. But $g_1 = \Phi_0$, and g_2, g_3 are product vectors distinct from, hence orthogonal to, Φ_0. Hence $0 = \langle\Phi_0|g_3\rangle = \frac{c_3\bar{c}_1}{\sqrt{2}}$, a contradiction. Thus $\hat{\alpha}(\varphi_0)$ is either φ_0 or φ_I.

APPENDIX I

Lemma. If $\mathfrak{A}$ is the CAR algebra in its Fock representation on Fock space $\mathcal{H}_{\mathcal{F}}$ and A is an operator in $\mathfrak{A}$ such that $A|\mathcal{H}_n = a(f_n)a(f_n)^*|\mathcal{H}_n$ for each n, then $A = a(f_1)a(f_1)^*$.

Proof. We shall show that $f_n = cf_m$ with $|c| = 1$; so that $a(f_n)a(f_n)^* = a(f_m)a(f_m)^*$ and $A = a(f_1)a(f_1)^*$. Suppose we have established that $f_0, f_1, \ldots, f_{m-1}$ differ from each other by scalar multiples of modulus 1. If $B = \Sigma c_{i_1 \ldots i_p, j_1 \ldots j_q} a(e_{i_1})^* \ldots a(e_{i_p})^* a(e_{j_1}) \ldots a(e_{j_q})$ and r is an integer larger than all the indices occurring in this sum, where $\{e_j\}$ is an orthonormal basis for $\mathcal{H}$ such that $f_1 = \|f_1\|e_1$ and $f_m \in [e_1, e_2]$, then

$$\|A-B\|^2 \geq \|(A-B)(e_1 \wedge e_{r+2} \wedge \ldots \wedge e_{r+m})\|^2 \geq |\,|\langle e_2|f_m\rangle|^2 - (c_{0;0} + c_{1;1})|^2$$

and

$$\|A-B\|^2 \geq \|(A-B)(e_1 \wedge e_{r+2} \wedge \dots \wedge e_{r+n})\|^2 \geq |c_{0;0} + c_{1;1}|^2 ,$$

when $n < m$; so that $|\langle e_2|f_m\rangle| \leq 2\|A-B\|$, for each such B; and $\langle e_2|f_m\rangle = 0$. Since $f_m \in [e_1, e_2]$, $f_m = ae_1 = a\|f_1\|^{-1}f_1$. Moreover

$$\|A-B\|^2 \geq \|(A-B)(e_2 \wedge e_{r+2} \wedge \dots \wedge e_{r+m})\|^2 \geq |\,|\langle e_1|f_m\rangle|^2 - (c_{0;0} + c_{2;2})|^2$$

and

$$\|A-B\|^2 \geq \|(A-B)(e_2 \wedge e_{r+2} \wedge \dots \wedge e_{r+n})\|^2 \geq |\,\|f_1\|^2 - (c_{0;0} + c_{2;2})|^2$$

when $n < m$; so that $|\,\|f_1\|^2 - |\langle e_1|f_m\rangle|^2| = |\,\|f_1\|^2 - |a|^2| \leq 2\|A-B\|$; and $|a| = \|f_1\|$. Thus $f_m = cf_1$, where $|c| = |a\|f_1\|^{-1}| = 1$.

APPENDIX II

Lemma. If $\mathfrak{A}$ is the CAR algebra in its Fock representation on Fock space $\mathcal{H}_{\mathcal{F}}$ and A is an operator in $\mathfrak{A}$ such that $A|\mathcal{H}_n = a(f_n)^*|\mathcal{H}_n$ for each n, then all f_n are equal (to f) and $A = a(f)^*$.

Proof. Suppose we have proved that $f_0 = f_1 = f_2 = \dots = f_{m-1}$ (=f). Let $\{e_j\}$ be an orthonormal basis for $\mathcal{H}$ such that $\|f\|^{-1}f = e_1$ and $f_m \in [e_1, e_2]$ (so that $f_m = \langle e_1|f_m\rangle e_1 + \langle e_2|f_m\rangle e_2$). If $B = \Sigma c_{i_1\dots i_p, j_1\dots j_q} a(e_{i_1})^* \dots a(e_{i_p})^* a(e_{j_1}) \dots a(e_{j_q})$ and r is an integer larger than any of the subscripts appearing in this finite sum then:

$$\|A-B\|^2 \geq \|(A-B)(e_{r+1} \wedge \dots \wedge e_{r+n})\|^2 = |c_{0;0}|^2 + |\,\|f\| - c_{1;0}|^2$$

$$\sum_{\{i_1\dots i_p\} \neq \{1\}} |c_{i_1\dots i_p;0}|^2$$

when $n < m$, and where a subscript '0' before the semicolon refers to the absence of creators and after the semicolon refers to the absence of annihilators ($c_{0;0}$ is the coefficient of I in the sum for B). We have, too,

$$\|A-B\|^2 \geq \|(A-B)(e_{r+1} \wedge \cdots \wedge e_{r+m})\|^2 = |c_{0;0}|^2 + |\langle e_1|f_m\rangle - c_{1;0}|^2$$

$$+ |\langle e_2|f_m\rangle - c_{2;0}|^2 + \sum_{\{i_1\ldots i_p\}\neq\{1\},\{2\}} |c_{i_1\ldots i_p;0}|^2 .$$

Thus

$$|\langle e_1|f_m\rangle - \|f\|| \leq 2\|A-B\|$$

and

$$|\langle e_2|f_m\rangle| \leq 2\|A-B\| .$$

Since B may be chosen so that $\|A-B\|$ is arbitrarily small, $\langle e_2|f_m\rangle = 0$. As $f_m \in [e_1, e_2]$, $f_m = ae_1$. In addition, $\|f\| = \langle e_1|f_m\rangle = a$. Thus $f_m = \|f\|e_1 = f$; and $A = a(f)^*$.

APPENDIX III

Proposition. If $\mathcal{V}$ and $\mathcal{W}$ are vector spaces, A and B are linear transformations of $\mathcal{V}$ into $\mathcal{W}$ such that for each v in $\mathcal{V}$ there is a scalar c_v for which $Bv = c_vAv$; then $B = cA$ for some scalar c.

Proof. Let $\mathcal{N}$ be the null space of A. From the hypothesized relation between A and B, $\mathcal{N}$ is contained in the null space of B. Thus A and B induce linear transformations $\bar{A}$ and $\bar{B}$ of the quotient space $\bar{\mathcal{V}}$ of $\mathcal{V}$ by $\mathcal{N}$ into $\mathcal{W}$ such that $A = \bar{A}\circ\eta$ and $B = \bar{B}\circ\eta$, where η is the quotient mapping of V onto $\overline{V}$. With v_0 in $\mathcal{N}$, $Bv_0 = c_{v_0} Av_0 = 0$; so that we may assume that $c_{v_0} = 0$ when $v_0 \in \mathcal{N}$. With this assumption, if $v \in V$ and $v_0 \in \mathcal{N}$, then $B(v+v_0) = c_{v+v_0} A(v+v_0) = c_{v+v_0} Av = Bv = c_v Av$. If $v \notin \mathcal{N}$ then $Av \neq 0$ so that $c_v = c_{v+v_0}$. If $v \in \mathcal{N}$ then $v + v_0 \in \mathcal{N}$ and $c_v = c_{v+v_0} = 0$. Thus, defining $c_{\overline{v}}$ to be c_v, for $\overline{v}$ in $\overline{V}$, where $\overline{v} = v + \mathcal{N}$, we have $\overline{B}\overline{v} = Bv = c_v Av = c_{\overline{v}} \overline{A}\overline{v}$. Note that the null space of $\overline{A}$ in $\overline{V}$ is (0). If we show that $\overline{B} = c\overline{A}$, for some scalar c then $Bv = \overline{B}\overline{v} = c\overline{A}\overline{v} = cAv$, for all v in V, so that $B = cA$. We may assume, from this discussion, that $\mathcal{N} = (0)$. With v and v' in V, we have $B(v+v') = c_{v+v'} A(v+v') = c_{v+v'} Av + c_{v+v'} Av' = Bv + Bv' = c_v Av + c_{v'} Av'$.

Thus $(c_{v+v'} - c_v) Av = (c_{v'} - c_{v+v'}) Av'$; and $c_v = c_{v+v'} = c_{v'}$, when v and v' are linearly independent. Let $\{v_a\}$ be a linear basis for V. Then $Bv_a = cAv_a$ for all a, where $c = c_{v_a}$ for all a. Thus $B = cA$.

Department of Mathematics E1
University of Pennsylvania
Philadelphia
Pennsylvania 19104
USA

Construction of ITPFI with non-trivial uncountable T-set

Motosige OSIKAWA

General Education Department, Kyushu University
Fukuoka 810, Japan

1. Associated flow and T-set

Let G be a countable group of null-measure preserving transformations of a Lebesgue space (Ω, P). We assume that T is ergodic. For g in G define a transformation $\tilde{g}$ of the product space $\Omega \times R$ by $\tilde{g}(\omega, u) = (g\omega, u - \log \frac{dPg}{dP}(\omega))$. We denote the set of all $\tilde{g}$ for g in G by $\tilde{G}$ and the measurable partition which generates the σ-algebra of all $\tilde{G}$-invariant sets by $\zeta(\tilde{G})$. Since the flow $\{T_s\}$ on $\Omega \times R$ defined by $T_s(\omega, u) = (\omega, u+s)$, $-\infty < s < +\infty$ commutes with $\tilde{G}$, $\{T_s\}$ induces a flow on the quotient space $X = \Omega \times R/\zeta(G)$. We denote it by the same $\{T_s\}$ and call it the associated flow of (Ω, P, G). The point spectra $S_p(\{T_s\})$ of a flow $\{T_s\}$ is the set of all real number t such that there exists a measurable function $\phi(x)$ on X with $|\phi(x)| = 1$ and $\phi(T_s x) = e^{its}\phi(x)$, a.e.x, $-\infty < s < +\infty$. A real number t is in the point spectra $S_p(\{T_s\})$ of the associated flow $\{T_s\}$ of (Ω, P, G) if and only if there exists a real measurable function $\psi(\omega)$ on Ω with $e^{it \log \frac{dPg}{dP}(\omega)} = e^{i\{\psi(g\omega) - \psi(\omega)\}}$, a.e.$\omega$, $g \in G$. The associated flow $\{T_s\}$ of (Ω, P, G) corresponds to the Takesaki dual action of the modular automorphism group of the group measure space construction factor of G restricted to the center and $S_p(\{T_s\})$ is the Connes T-set of one.

2. Definition of AC-flow

For $k = 1,2,\ldots$ let Q_k be a probability measure on l_k point set $X_k = \{1,2,\ldots,l_k\}$. Let (X, Q) be the infinite direct product space of (X_k, Q_k), $k = 1,2,\ldots$. For an element x in X, x_k denotes the k-th coordinate of x. An adding machine T is the transformation of X defined as follows; $(Tx)_j = 1$ $(j = 1,2,\ldots,k-1)$, $= x_k+1$ $(j = k)$, $=x_j$ $(j = k+1,\ldots)$ if $x_j = l_j$ $(j = 1,2,\ldots,k-1)$ and $x_k \neq l_k$. Let

$_k(x_k)$ be a function on X_k such that $\xi_k(1) = 0$ and $\xi_k(x_k+1) >$ $_k(x_k) + \sum_{j=1}^{k-1}\xi_j(1_j)$ if $x_k \neq 1_k$, $k = 1,2,\ldots.$ We may consider $\{\xi_k\}$ s a sequence of independent variables on X. We define a positive alued function $f(x)$ on X by $f(x) = \xi_k(x_k+1) - \xi_k(x_k) - \sum_{j=1}^{k-1}\xi_j(1_j)$ f $x_j = 1_j$ $(j = 1,2,\ldots,k-1)$ and $x_k \neq 1_k$, $k = 1,2,\ldots.$ The flow $T_s\}$ built under a ceiling function $f(x)$ based on an adding machine is called the AC-flow generated by (X_k, Q_k, ξ_k), $k = 1,2,\ldots.$ For uch an AC-flow $\{T_s\}$ a real number t is in $S_p(\{T_s\})$ if and only if here exists a sequence $\{c_j\}$ of real numbers such that $e^{it\sum_{j=1}^{n}(X_j(x)-c_j)}$ onverges a.e. as $n \to \infty$.

EXAMPLE. Let $M_1 = 2$, $M_k = 2^k M_{k-1}$ $(k = 2,3,\ldots)$, $X_k = \{1,2\}$, $_k = \{\frac{\lambda}{1+\lambda}, \frac{1}{1+\lambda}\}$ for fixed λ, $\xi_k(1) = 0$ and $\xi_k(2) = M_k$ $(k = 1,2,\ldots)$. hen we have

$$_p(\{T_s\}) = \frac{1_1}{M_1} + \frac{1_2}{M_2} + \ldots + \frac{1_k}{M_k} + \ldots\ ;\ 1_k\text{'s are integers and } \sum_{k=1}^{\infty}\left(\frac{1_k}{2^k}\right)^2 < +\infty\}.$$

he last set is a non-trivial uncountable subgroup of real numbers.

3. ITPFI and AC-flow

Let $\{q_{k,i}\ ;\ i = 1,2,\ldots,1_k,\ k = 1,2,\ldots\}$ and $\{m_{k,i}\ ;\ i = 1,2,\ldots,$ $_k$, $k = 1,2,\ldots\}$ be a sequences of positive numbers and positive integers espectively which satisfy

$$\frac{q_{k,i+1}}{q_{k,i}} > \prod_{j=1}^{k-1}\frac{q_{j,1_j}}{q_{j,1}} \quad (i = 1,2,\ldots,1_k,\ k = 2,3,\ldots) \quad \text{and}$$

$$m_{k,1}q_{k,1} + \ldots + m_{k,1_k}q_{k,1_k} = 1 \quad (k = 1,2,\ldots).$$

Put $n_k = m_{k,1} + \ldots + m_{k,1_k}$ $(k = 1,2,\ldots)$ and let P_k be the ollowing probability measure on n_k point space $\Omega_k = \{1,2,\ldots,n_k\}$;

$$_k(\{r\}) = q_{k,1} \quad \text{for} \quad \sum_{j=1}^{i-1}m_{k,j}+1 \leq r \leq \sum_{j=1}^{i}m_{k,j}\ ,\ (i = 1,2,\ldots 1_k,\ k = 1,2,$$

..). Let (Ω, P) be the infinite direct product space of (Ω_k, P_k), $= 1,2,\ldots.$ We may consider that the permutation group G_k of the Ω_k a transformation group of (Ω, P). We denote by G the group nerated by $\bigcup_{k=1}^{\infty} G_k$. The group measure space construction factor conructed from (Ω, P, G) is ITPFI. The associated flow of (Ω, P, G)

is the AC-flow generated by the following (X_k, Q_k, ξ_k); $X_k = \{1,2,\ldots,l_k\}$, $Q_k(\{i\}) = m_{k,i}q_{k,i}$, $\xi_k(i) = \log\frac{q_{k,i}}{q_{k,1}}$. Furthermore, for any AC-flow ther exists such (Ω, P, G) with it as the associated flow.

REFERENCES

[1] A. Connes, On the hierarchy of W. Krieger, Illinois J. Math. 19 (1975) 428-432.

[2] T. Hamachi, Y. Oka and M. Osikawa, A classification of ergodic non-singular transformation groups, Mem. Fac. Sci. Kyushu Univ. 18 (1974) 113-133.

[3] T. Hamaci, Y. Oka and M. Osikawa, Flows associated with ergodic non-singular transformation groups, Publ. RIMS, Kyoto Univ. 11 (1975) 31-50.

[4] M. Osikawa, Point spectra of non-singular flows, to appear in Publ. RIMS Kyoto Univ.

[5] E. J. Woods, The classification of factors is not smooth, Can. J. Math. 15 (1973) 96-102.

ON THE ALGEBRAIC REDUCTION THEORY FOR COUNTABLE DIRECT SUMMAND C*-ALGEBRAS OF SEPARABLE C*-ALGEBRAS

Hideo TAKEMOTO
College of General Education
Tohoku University
Sendai, Japan

In this paper, we shall show that certain quotient algebras of a *-algebra are von Neumann algebras. More specifically if $\mathfrak{A}$ is a C*-lgebra represented as the countable direct sum of a separable C*-algera, then we show that many quotient algebras of $\mathfrak{A}$ are von Neumann lgebras. This result is, in a sense, an extension of the results in the ase of finite von Neumann algebras by Sakai [1], Takesaki [3] and Takeoto and Tomiyama [2]. In the case of finite von Neumann algebras, we ave the following result by Takesaki [3]:

Let $\mathfrak{A}$ be a finite von Neumann algebra and $\mathcal{A}$ a von Neumann ubalgebra of the center of $\mathfrak{A}$. Further, let Ω be the spectrum space. uppose that there exists a faithfull normal expectation ε from $\mathfrak{A}$ nto $\mathcal{A}$ satisfying $\varepsilon(x^*x) = \varepsilon(xx^*)$ for all $x \in \mathfrak{A}$. For every $\in \Omega$, put $\mathfrak{m}_\omega = \{ a \in \mathfrak{A} \ ; \ \varepsilon(a^*a)(\omega) = 0 \}$, then the quotient algebra $\mathfrak{A}/\mathfrak{m}_\omega$ becomes a von Neumann algebra. Furthermore, let π_ω be the anonical homomorphism of $\mathfrak{A}$ onto $\mathfrak{A}/\mathfrak{m}_\omega$ and $\mathcal{B}$ a von Neumann sub-lgebra of $\mathfrak{A}$ containing $\mathcal{A}$. Then, $\pi_\omega(\mathcal{B})$ is a von Neumann subal-ebra of $\mathfrak{A}/\mathfrak{m}_\omega$.

Takesaki's result is an extension of Sakai's result in which the lgebra $\mathcal{A}$ is the center of original algebra. Furthermore, Takemoto nd Tomiyama [2] reproved the above Takesaki's result by considering the ontinuous field V of which each fibre is a linear functional on $\mathfrak{A}/\mathfrak{m}_\omega$

In the proof of [2] the following result was shown:

Let $\mathcal{B}$ be a C*-subalgebra of $\mathfrak{A}$ containing $\mathcal{A}$ and ω_0 an element of Ω, then $\widetilde{\pi_{\omega_0}(\mathcal{B})}$ coincides with $\pi_{\omega_0}(\widetilde{\mathcal{B}})$ if and only if, for every $\Phi \in V$, the function $\omega \to \|\Phi(\omega) \mid \pi(\mathcal{B})\|$ is continuous at ω_0.

From a point of view of continuous fields one has the following problem. Let N be the set of all positive integers and βN the Stone-Cech compactfication of N. Let $\mathcal{A}$ be a separable C*-algebra with identity and Φ^0 a faithfull state of $\mathcal{A}$. Let $\mathfrak{A}$ be the C*-algebra $l^\infty(N,\mathcal{A}) = \{a = (\bar{a}_n);$ bounded sequences in $\mathcal{A}\}$. Then, for every $\omega \in \beta N$, we can define a closed two-sided ideal $\mathfrak{M}_\omega$ by:

$$\mathfrak{M}_\omega = \{\, a = (\bar{a}_n);\ \lim \Phi^0(\bar{b}_n\bar{a}_n\bar{c}_n) = 0 \quad \text{for every } b,\ c, \in \mathfrak{A} \,\}$$

where $\mathfrak{U}$ is a filter consisting of allneighborhoods of ω. Now, if ω is an element of N, the quotient algebra $\mathfrak{A}/\mathfrak{M}_\omega$ is isomorphic to the algebra $\mathcal{A}$ so that $\mathfrak{A}/\mathfrak{M}_\omega$ does not become a von Neumann algebra From the above considerations, we have the problem of whether, for every $\omega \in \beta N\setminus N$, the quotient algebra $\mathfrak{A}/\mathfrak{M}_\omega$ is a von Neumann algebra.

In the following cases, we can settle this problem.

(1) If $\mathcal{A}$ is a non-separable C*-algebra, the weak closure $\widetilde{\mathcal{A}}$ is a finite von Neumann algebra and Φ^0 is a trace of $\mathcal{A}$ .By using Takesaki's result, we can get the same result in this case

(2) If $\mathcal{A}$ is separable in the uniform topology and is a von Neumann algebra, $\mathcal{A}$ is finite dimensional. Thus, the above problem as an affirmative answer.

For the above problem, we may assume that the C*-algebra $\mathcal{A}$ is acting on a Hilbert space H and there is a cyclic vector ξ_0 in H with $\phi^0(\bar{a}) = (\bar{a}\xi_0|\xi_0)$ for every $\bar{a} \in \mathcal{A}$. For every a and b of $\mathfrak{A}$, define a $l^\infty(N,C)$-module homomorphism $R_aL_b\Phi^0 = (R_{\bar{a}_n}L_{\bar{b}_n}\phi^0)$ of $\mathfrak{A}$ to $l^\infty(N,C)$. Then $R_aL_b\Phi^0$ is bounded. We let V be the closure in $B(\mathfrak{A}, l^\infty(N,C))$ of the set $\{R_aL_b\Phi^0;\ a,b \in \mathfrak{A}\}$ where $B(\mathfrak{A}, l^\infty(N,C)$ is the Banach space consisting of all bounded operators of $\mathfrak{A}$ to $l^\infty(N,C)$. Let W be the closure in $\mathcal{A}^*$ of $\{R_{\bar{a}}L_{\bar{b}}\phi^0;\ \bar{a},\bar{b} \in \mathcal{A}\}$. Then every element ϕ of W extends uniquely to a normal functional of $\widetilde{\mathcal{A}}$, where $\widetilde{\mathcal{A}}$ is the weak closure of $\mathcal{A}$. Furthermore, for every $\Phi \in V$, $\Phi(\mathfrak{M}_\omega) \to \{0\}$. Let $\Phi_\omega(\tilde{a}) = \lim_{\mathfrak{U}} \Phi_n(\bar{a}_n)$ for $\Phi = (\Phi_n)$ and $a \in \mathfrak{A}$, Then denotes Φ_ω is a bounded linear functional on $\mathfrak{A}/\mathfrak{M}_\omega$, where $\tilde{a}$ denotes the class of a in $\mathfrak{A}/\mathfrak{M}_\omega$ and $\Phi = (\Phi_n)$ is the representation of Φ determined in Lemma 1 (below). With $V_\omega = \{\Phi_\omega;\ \Phi \in V\}$, one can show that V_ω is a closed invariant subspace in $(\mathfrak{A}/\mathfrak{M}_\omega)^*$.

With the above notation, we have following theorem.

THEOREM. Let $\mathcal{A}$ be a separable C*-algebra with identity and ϕ^0 a faithful state of $\mathcal{A}$. Let $\mathfrak{A} = l^\infty(N,\mathcal{A})$ and, for every $\omega \in \beta N\setminus N$, $\mathfrak{U}$ be the filter consisting of all neighborhoods of ω, and

$\mathfrak{M}_\omega = \{a \in \mathfrak{A}\ ;\ \lim_{\mathfrak{U}} \phi^{\circ}(\bar{b}_n\bar{a}_n\bar{c}_n) = 0$ for every $b, c \in \mathfrak{A}\}$. Then, the quotient algebra $\mathfrak{A}/\mathfrak{M}_\omega$ is a von Neumann algebra.

We prove the above theorem with the following lemmas.

LEMMA 1. For every $\Phi \in V$, there exists a bounded sequence (ϕ_n) in W satisfying $\Phi(a) = (\phi_n(\bar{a}_n))$ for every $a = (\bar{a}_n) \in \mathfrak{A}$ and $\|\Phi\| = \sup_n \|\phi_n\|$.

By the assumption on $\mathfrak{A}$, each ϕ of W has a unique normal extension $\tilde{\phi}$ to $\tilde{\mathfrak{A}}$. Thus, we have the following lemma.

LEMMA 2. The subspace V_ω of $(\mathfrak{A}/\mathfrak{M}_\omega)^*$ is a closed invariant subspace.

By Lemma 2, $(\mathfrak{A}/\mathfrak{M}_\omega)^{**}/V_\omega$ is a von Neumann algebra with predual V_ω. We can show that $\mathfrak{A}/\mathfrak{M}_\omega$ is canonically embeded in $(\mathfrak{A}/\mathfrak{M}_\omega)^{**}/V_\omega$ and is weakly dense there.

Next, let $\{\psi^{(\ell)}\}$ be a countable dense sub set of the unit ball of W and $\Psi^{(\ell)}$ the constant field with fibre $\psi^{(\ell)}$. Then, considering the absolute value $|\psi^{(\ell)}|$ of $\psi^{(\ell)}$, we can define $|\Psi_\omega^{(\ell)}|(a) = \lim_{\mathfrak{U}} |\psi^{(\ell)}|(a_n)$ for every a. Thus we have;

LEMMA 3. For an arbitrary self-adjoint element A of $(\mathfrak{A}/\mathfrak{M}_\omega)^{**}/V_\omega$, there exists a sequence $\{a^{(k)}\}_{k=1}^{\infty}$ of self-adjoint elements in $\mathfrak{A}$ satisfying the following conditions:

(1) $|\Psi_\omega^{(\ell)}|((A - \tilde{a}^{(k)})^2) < (1/2^{k+1})^2$ for $1 \leq k$,

(2) $\lim_{\mathfrak{U}} |\psi^{(\ell)}|((a_n^{(k)} - a_n^{(k+1)})^2) < (1/2^{k})^2$ for $\leq k$

REFERENCES

[1] S. Sakai, Lecture Note, Yale University, 1962.

[2] H. Takemoto and J. Tomiyama, On the topological reduction of finite von Neumann algebras, Tôhoku Math. J., 25(1973), 273-289.

[3] M. Takesaki, The quotient algebra of a finite von Neumann algebra, Pacific J. Math., 36(1971), 827-831.

ol. 489: J. Bair and R. Fourneau, Etude Géométrique des Espaces ectoriels. Une Introduction. VII, 185 pages. 1975.

ol. 490: The Geometry of Metric and Linear Spaces. Proceedings 974. Edited by L. M. Kelly. X, 244 pages. 1975.

ol. 491: K. A. Broughan, Invariants for Real-Generated Uniform opological and Algebraic Categories. X, 197 pages. 1975.

ol. 492: Infinitary Logic: In Memoriam Carol Karp. Edited by D. W. eker. VI, 206 pages. 1975.

ol. 493: F. W. Kamber and P. Tondeur, Foliated Bundles and haracteristic Classes. XIII, 208 pages. 1975.

ol. 494: A Cornea and G. Licea. Order and Potential Resolvent amilies of Kernels. IV, 154 pages. 1975.

ol. 495: A. Kerber, Representations of Permutation Groups II. V, 5 pages. 1975.

ol. 496: L. H. Hodgkin and V. P. Snaith, Topics in K-Theory. Two dependent Contributions. III, 294 pages. 1975.

ol. 497: Analyse Harmonique sur les Groupes de Lie. Proceedings 973–75. Edité par P. Eymard et al. VI, 710 pages. 1975.

ol. 498: Model Theory and Algebra. A Memorial Tribute to braham Robinson. Edited by D. H. Saracino and V. B. Weispfenning. 463 pages. 1975.

ol. 499: Logic Conference, Kiel 1974. Proceedings. Edited by . H. Müller, A. Oberschelp, and K. Potthoff. V, 651 pages 1975.

ol. 500: Proof Theory Symposion, Kiel 1974. Proceedings. Edited by Diller and G. H. Müller. VIII, 383 pages. 1975.

ol. 501: Spline Functions, Karlsruhe 1975. Proceedings. Edited by Böhmer, G. Meinardus, and W. Schempp. VI, 421 pages. 1976.

ol. 502: János Galambos, Representations of Real Numbers by finite Series. VI, 146 pages. 1976.

ol. 503: Applications of Methods of Functional Analysis to Problems Mechanics. Proceedings 1975. Edited by P. Germain and B. ayroles. XIX, 531 pages. 1976.

ol. 504: S. Lang and H. F. Trotter, Frobenius Distributions in L_2-Extensions. III, 274 pages. 1976.

ol. 505: Advances in Complex Function Theory. Proceedings 973/74. Edited by W. E. Kirwan and L. Zalcman. VIII, 203 pages. 976.

ol. 506: Numerical Analysis, Dundee 1975. Proceedings. Edited y G. A. Watson. X, 201 pages. 1976.

ol. 507: M. C. Reed, Abstract Non-Linear Wave Equations. VI, 28 pages. 1976.

ol. 508: E. Seneta, Regularly Varying Functions. V, 112 pages. 1976.

ol. 509: D. E. Blair, Contact Manifolds in Riemannian Geometry. I, 146 pages. 1976.

ol. 510: V. Poènaru, Singularités C^∞ en Présence de Symétrie. , 174 pages. 1976.

ol. 511: Séminaire de Probabilités X. Proceedings 1974/75. Edité ar P. A. Meyer. VI, 593 pages. 1976.

ol. 512: Spaces of Analytic Functions, Kristiansand, Norway 1975. roceedings. Edited by O. B. Bekken, B. K. Øksendal, and A. Stray. II, 204 pages. 1976.

ol. 513: R. B. Warfield, Jr. Nilpotent Groups. VIII, 115 pages. 1976.

Vol. 514: Séminaire Bourbaki vol. 1974/75. Exposés 453 – 470. IV, 276 pages. 1976.

ol. 515: Bäcklund Transformations. Nashville, Tennessee 1974. Proceedings. Edited by R. M. Miura. VIII, 295 pages. 1976.

ol. 516: M. L. Silverstein, Boundary Theory for Symmetric Markov rocesses. XVI, 314 pages. 1976.

ol. 517: S. Glasner, Proximal Flows. VIII, 153 pages. 1976.

ol. 518: Séminaire de Théorie du Potentiel, Proceedings Paris 972–1974. Edité par F. Hirsch et G. Mokobodzki. VI, 275 pages. 976.

ol. 519: J. Schmets, Espaces de Fonctions Continues. XII, 150 ages. 1976.

ol. 520: R. H. Farrell, Techniques of Multivariate Calculation. X, 37 pages. 1976.

Vol. 521: G. Cherlin, Model Theoretic Algebra – Selected Topics. IV, 234 pages. 1976.

Vol. 522: C. O. Bloom and N. D. Kazarinoff, Short Wave Radiation Problems in Inhomogeneous Media: Asymptotic Solutions. V. 104 pages. 1976.

Vol. 523: S. A. Albeverio and R. J. Høegh-Krohn, Mathematical Theory of Feynman Path Integrals. IV, 139 pages. 1976.

Vol. 524: Séminaire Pierre Lelong (Analyse) Année 1974/75. Edité par P. Lelong. V, 222 pages. 1976.

Vol. 525: Structural Stability, the Theory of Catastrophes, and Applications in the Sciences. Proceedings 1975. Edited by P. Hilton. VI, 408 pages. 1976.

Vol. 526: Probability in Banach Spaces. Proceedings 1975. Edited by A. Beck. VI, 290 pages. 1976.

Vol. 527: M. Denker, Ch. Grillenberger, and K. Sigmund, Ergodic Theory on Compact Spaces. IV, 360 pages. 1976.

Vol. 528: J. E. Humphreys, Ordinary and Modular Representations of Chevalley Groups. III, 127 pages. 1976.

Vol. 529: J. Grandell, Doubly Stochastic Poisson Processes. X, 234 pages. 1976.

Vol. 530: S. S. Gelbart, Weil's Representation and the Spectrum of the Metaplectic Group. VII, 140 pages. 1976.

Vol. 531: Y.-C. Wong, The Topology of Uniform Convergence on Order-Bounded Sets. VI, 163 pages. 1976.

Vol. 532: Théorie Ergodique. Proceedings 1973/1974. Edité par J.-P. Conze and M. S. Keane. VIII, 227 pages. 1976.

Vol. 533: F. R. Cohen, T. J. Lada, and J. P. May, The Homology of Iterated Loop Spaces. IX, 490 pages. 1976.

Vol. 534: C. Preston, Random Fields. V, 200 pages. 1976.

Vol. 535: Singularités d'Applications Differentiables. Plans-sur-Bex. 1975. Edité par O. Burlet et F. Ronga. V, 253 pages. 1976.

Vol. 536: W. M. Schmidt, Equations over Finite Fields. An Elementary Approach. IX, 267 pages. 1976.

Vol. 537: Set Theory and Hierarchy Theory. Bierutowice, Poland 1975. A Memorial Tribute to Andrzej Mostowski. Edited by W. Marek, M. Srebrny and A. Zarach. XIII, 345 pages. 1976.

Vol. 538: G. Fischer, Complex Analytic Geometry. VII, 201 pages. 1976.

Vol. 539: A. Badrikian, J. F. C. Kingman et J. Kuelbs, Ecole d'Eté de Probabilités de Saint Flour V-1975. Edité par P.-L. Hennequin. IX, 314 pages. 1976.

Vol. 540: Categorical Topology, Proceedings 1975. Edited by E. Binz and H. Herrlich. XV, 719 pages. 1976.

Vol. 541: Measure Theory, Oberwolfach 1975. Proceedings. Edited by A. Bellow and D. Kölzow. XIV, 430 pages. 1976.

Vol. 542: D. A. Edwards and H. M. Hastings, Čech and Steenrod Homotopy Theories with Applications to Geometric Topology. VII, 296 pages. 1976.

Vol. 543: Nonlinear Operators and the Calculus of Variations, Bruxelles 1975. Edited by J. P. Gossez, E. J. Lami Dozo, J. Mawhin, and L. Waelbroeck, VII, 237 pages. 1976.

Vol. 544: Robert P. Langlands, On the Functional Equations Satisfied by Eisenstein Series. VII, 337 pages. 1976.

Vol. 545: Noncommutative Ring Theory. Kent State 1975. Edited by J. H. Cozzens and F. L. Sandomierski. V, 212 pages. 1976.

Vol. 546: K. Mahler, Lectures on Transcendental Numbers. Edited and Completed by B. Diviš and W. J. Le Veque. XXI, 254 pages. 1976.

Vol. 547: A. Mukherjea and N. A. Tserpes, Measures on Topological Semigroups: Convolution Products and Random Walks. V, 197 pages. 1976.

Vol. 548: D. A. Hejhal, The Selberg Trace Formula for PSL (2,IR). Volume I. VI, 516 pages. 1976.

Vol. 549: Brauer Groups, Evanston 1975. Proceedings. Edited by D. Zelinsky. V, 187 pages. 1976.

Vol. 550: Proceedings of the Third Japan – USSR Symposium on Probability Theory. Edited by G. Maruyama and J. V. Prokhorov. VI, 722 pages. 1976.

Vol. 551: Algebraic K-Theory, Evanston 1976. Proceedings. Edited by M. R. Stein. XI, 409 pages. 1976.

Vol. 552: C. G. Gibson, K. Wirthmüller, A. A. du Plessis and E. J. N. Looijenga. Topological Stability of Smooth Mappings. V, 155 pages. 1976.

Vol. 553: M. Petrich, Categories of Algebraic Systems. Vector and Projective Spaces, Semigroups, Rings and Lattices. VIII, 217 pages. 1976.

Vol. 554: J. D. H. Smith, Mal'cev Varieties. VIII, 158 pages. 1976.

Vol. 555: M. Ishida, The Genus Fields of Algebraic Number Fields. VII, 116 pages. 1976.

Vol. 556: Approximation Theory. Bonn 1976. Proceedings. Edited by R. Schaback and K. Scherer. VII, 466 pages. 1976.

Vol. 557: W. Iberkleid and T. Petrie, Smooth S^1 Manifolds. III, 163 pages. 1976.

Vol. 558: B. Weisfeiler, On Construction and Identification of Graphs. XIV, 237 pages. 1976.

Vol. 559: J.-P. Caubet, Le Mouvement Brownien Relativiste. IX, 212 pages. 1976.

Vol. 560: Combinatorial Mathematics, IV, Proceedings 1975. Edited by L. R. A. Casse and W. D. Wallis. VII, 249 pages. 1976.

Vol. 561: Function Theoretic Methods for Partial Differential Equations. Darmstadt 1976. Proceedings. Edited by V. E. Meister, N. Weck and W. L. Wendland. XVIII, 520 pages. 1976.

Vol. 562: R. W. Goodman, Nilpotent Lie Groups: Structure and Applications to Analysis. X, 210 pages. 1976.

Vol. 563: Séminaire de Théorie du Potentiel. Paris, No. 2. Proceedings 1975–1976. Edited by F. Hirsch and G. Mokobodzki. VI, 292 pages. 1976.

Vol. 564: Ordinary and Partial Differential Equations, Dundee 1976. Proceedings. Edited by W. N. Everitt and B. D. Sleeman. XVIII, 551 pages. 1976.

Vol. 565: Turbulence and Navier Stokes Equations. Proceedings 1975. Edited by R. Temam. IX, 194 pages. 1976.

Vol. 566: Empirical Distributions and Processes. Oberwolfach 1976. Proceedings. Edited by P. Gaenssler and P. Révész. VII, 146 pages. 1976.

Vol. 567: Séminaire Bourbaki vol. 1975/76. Exposés 471–488. IV, 303 pages. 1977.

Vol. 568: R. E. Gaines and J. L. Mawhin, Coincidence Degree, and Nonlinear Differential Equations. V, 262 pages. 1977.

Vol. 569: Cohomologie Etale SGA 4½. Séminaire de Géométrie Algébrique du Bois-Marie. Edité par P. Deligne. V, 312 pages. 1977.

Vol. 570: Differential Geometrical Methods in Mathematical Physics, Bonn 1975. Proceedings. Edited by K. Bleuler and A. Reetz. VIII, 576 pages. 1977.

Vol. 571: Constructive Theory of Functions of Several Variables, Oberwolfach 1976. Proceedings. Edited by W. Schempp and K. Zeller. VI. 290 pages. 1977

Vol. 572: Sparse Matrix Techniques, Copenhagen 1976. Edited by V. A. Barker. V, 184 pages. 1977.

Vol. 573: Group Theory, Canberra 1975. Proceedings. Edited by R. A. Bryce, J. Cossey and M. F. Newman. VII, 146 pages. 1977.

Vol. 574: J. Moldestad, Computations in Higher Types. IV, 203 pages. 1977.

Vol. 575: K-Theory and Operator Algebras, Athens, Georgia 1975. Edited by B. B. Morrel and I. M. Singer. VI, 191 pages. 1977.

Vol. 576: V. S. Varadarajan, Harmonic Analysis on Real Reductive Groups. VI, 521 pages. 1977.

Vol. 577: J. P. May, E_∞ Ring Spaces and E_∞ Ring Spectra. IV, 268 pages. 1977.

Vol. 578: Séminaire Pierre Lelong (Analyse) Année 1975/76. Edité par P. Lelong. VI, 327 pages. 1977.

Vol. 579: Combinatoire et Représentation du Groupe Symétrique, Strasbourg 1976. Proceedings 1976. Edité par D. Foata. IV, 339 pages. 1977.

Vol. 580: C. Castaing and M. Valadier, Convex Analysis and Measurable Multifunctions. VIII, 278 pages. 1977.

Vol. 581: Séminaire de Probabilités XI, Université de Strasbourg. Proceedings 1975/1976. Edité par C. Dellacherie, P. A. Meyer et M. Weil. VI, 574 pages. 1977.

Vol. 582: J. M. G. Fell, Induced Representations and Banach *-Algebraic Bundles. IV, 349 pages. 1977.

Vol. 583: W. Hirsch, C. C. Pugh and M. Shub, Invariant Manifolds. IV, 149 pages. 1977.

Vol. 584: C. Brezinski, Accélération de la Convergence en Analyse Numérique. IV, 313 pages. 1977.

Vol. 585: T. A. Springer, Invariant Theory. VI, 112 pages. 1977.

Vol. 586: Séminaire d'Algèbre Paul Dubreil, Paris 1975–1976 (29ème Année). Edited by M. P. Malliavin. VI, 188 pages. 1977.

Vol. 587: Non-Commutative Harmonic Analysis. Proceedings 1976. Edited by J. Carmona and M. Vergne. IV, 240 pages. 1977.

Vol. 588: P. Molino, Théorie des G-Structures: Le Problème d'Equivalence. VI, 163 pages. 1977.

Vol. 589: Cohomologie l-adique et Fonctions L. Séminaire de Géométrie Algébrique du Bois-Marie 1965–66, SGA 5. Edité par L. Illusie. XII, 484 pages. 1977.

Vol. 590: H. Matsumoto, Analyse Harmonique dans les Systèmes de Tits Bornologiques de Type Affine. IV, 219 pages. 1977.

Vol. 591: G. A. Anderson, Surgery with Coefficients. VIII, 157 pages. 1977.

Vol. 592: D. Voigt, Induzierte Darstellungen in der Theorie der endlichen, algebraischen Gruppen. V, 413 Seiten. 1977.

Vol. 593: K. Barbey and H. König, Abstract Analytic Function Theory and Hardy Algebras. VIII, 260 pages. 1977.

Vol. 594: Singular Perturbations and Boundary Layer Theory, Lyon 1976. Edited by C. M. Brauner, B. Gay, and J. Mathieu. VIII, 539 pages. 1977.

Vol. 595: W. Hazod, Stetige Faltungshalbgruppen von Wahrscheinlichkeitsmaßen und erzeugende Distributionen. XIII, 157 Seiten. 1977.

Vol. 596: K. Deimling, Ordinary Differential Equations in Banach Spaces. VI, 137 pages. 1977.

Vol. 597: Geometry and Topology, Rio de Janeiro, July 1976. Proceedings. Edited by J. Palis and M. do Carmo. VI, 866 pages. 1977.

Vol. 598: J. Hoffmann-Jørgensen, T. M. Liggett et J. Neveu, Ecole d'Eté de Probabilités de Saint-Flour VI – 1976. Edité par P.-L. Hennequin. XII, 447 pages. 1977.

Vol. 599: Complex Analysis, Kentucky 1976. Proceedings. Edited by J. D. Buckholtz and T. J. Suffridge. X, 159 pages. 1977.

Vol. 600: W. Stoll, Value Distribution on Parabolic Spaces. VIII, 216 pages. 1977.

Vol. 601: Modular Functions of one Variable V, Bonn 1976. Proceedings. Edited by J.-P. Serre and D. B. Zagier. VI, 294 pages. 1977.

Vol. 602: J. P. Brezin, Harmonic Analysis on Compact Solvmanifolds. VIII, 179 pages. 1977.

Vol. 603: B. Moishezon, Complex Surfaces and Connected Sums of Complex Projective Planes. IV, 234 pages. 1977.

Vol. 604: Banach Spaces of Analytic Functions, Kent, Ohio 1976. Proceedings. Edited by J. Baker, C. Cleaver and Joseph Diestel. VI, 141 pages. 1977.

Vol. 605: Sario et al., Classification Theory of Riemannian Manifolds. XX, 498 pages. 1977.

Vol. 606: Mathematical Aspects of Finite Element Methods. Proceedings 1975. Edited by I. Galligani and E. Magenes. VI, 362 pages. 1977.

Vol. 607: M. Métivier, Reelle und Vektorwertige Quasimartingale und die Theorie der Stochastischen Integration. X, 310 Seiten. 1977.

Vol. 608: Bigard et al., Groupes et Anneaux Réticulés. XIV, 334 pages. 1977.